ON LAUGHTER-SILVERED WINGS

This book is
dedicated to the memory of
Nurse Penny Clarke
and
South African Air Force pilot Major Frank Robertson

ON LAUGHTER-SILVERED WINGS

*The Story of
Lt. Col. E.T. (Ted) Strever DFC*

(who, in 1942, with his air crew carried out
the first mid-air skyjack in history)

Gail Strever-Morkel

Pen & Sword
AVIATION

First published in Great Britain
and reprinted in this format in 2022 by
PEN AND SWORD AVIATION
an imprint of
Pen and Sword Books Ltd
47 Church Street
Barnsley
South Yorkshire S70 2AS

Copyright © Gail Strever-Morkel, 2013, 2022

ISBN 978-1-39907-498-8

The right of Gail Strever-Morkel to be identified
as the author of this work has been asserted by her
in accordance with the Copyright, Designs and Patents Act 1988.

A CIP record for this book is available from the British Library.

All rights reserved. No part of this book may be reproduced or transmitted
in any form or by any means, electronic or mechanical including photocopying,
recording or by any information storage and retrieval
system, without permission from the Publisher in writing.

Printed and bound in England by
CPI Group (UK) Ltd, Croydon, CR0 4YY

Typeset in Times by CHIC GRAPHICS

Pen & Sword Books Ltd incorporates the imprints of
Pen & Sword Aviation, Pen & Sword Family History, Pen & Sword Maritime,
Pen & Sword Military, Pen & Sword Discovery, Wharncliffe Local History,
Wharncliffe True Crime, Wharncliffe Transport, Pen & Sword Select,
Pen & Sword Military Classics, Leo Cooper, Remember When,
The Praetorian Press, Seaforth Publishing and Frontline Publishing

For a complete list of Pen and Sword titles please contact
Pen and Sword Books Limited
47 Church Street, Barnsley, South Yorkshire, S70 2AS, England
E-mail: enquiries@pen-and-sword.co.uk
Website: www.pen-and-sword.co.uk

Contents

Acknowledgements vii

Prologue x

Chapter 1   In at the Deep End 1

Chapter 2   Ignorance is Bliss? 12

Chapter 3   From Whence He Came 23

Chapter 4   A Shifting Permanence 36

Chapter 5   Like Father, Like Son 40

Chapter 6   Turning Point 50

Chapter 7   Bucking the System 59

Chapter 8   Taking Flight 65

Chapter 9   Cloud Nine 75

Chapter 10   Taking the Controls 81

Chapter 11   The Big, Wide World 87

Chapter 12   Puttin' on the Ritz 96

Chapter 13   Dive-Bombing and Boyish Pranks 100

ON LAUGHTER-SILVERED WINGS

Chapter 14   Where There's a Will, There's a Way! 113

Chapter 15   Doing the Flying Cha-Cha-Cha 130

Chapter 16   Thumb Twiddling 153

Chapter 17   Bombay and Biding Time 160

Chapter 18   A Flamer – As Fate Would Have It! 167

Chapter 19   The Soft Touch of Home 176

Chapter 20   Burma: 'With Speed to the Stars'? 183

Epilogue 200

Appendix 213

Notes 216

Bibliography & Sources 220

Index 220

Acknowledgements

Darling Dad, thank you for at last letting me fill my blank-paged book with a part of your life's story.

I feel immensely humbled and privileged to have been the recipient of personal, handwritten letters from so many. Most of my respondents were in their eighties; the health of some was seriously failing and I know that their detailed and invaluable replies to my queries were accomplished with great effort. I am deeply grateful, and painfully aware that most are no longer alive, and that what I hold of them has become all the more precious. My thanks go to interviewees Andrée Strever and Wally Levy, who gave me permission to use their interview material in the book. Thanks to Hugh J. Wattington, S.J. Haffenden, Ray Brown, Jinx Glennie-Carr and Riccardo Alessio, who gave me permission to use their responses to my questionnaires, and for their correspondence. Thanks for the permission granted by Bob Rogers, Herschell Reilley and Eddie Whiston to reproduce their writing; all, on receiving portions of the transcripts of Dad's tapes in which they featured, gladly filled in the gaps by writing comprehensive additions from their memories. And thanks to Carey Heydenrych for giving me a copy of his own book as well as material that had been edited out of it, and the permission to use them along with my interview with him to fill in the gaps in the manuscript.

They were the men and women of a rapidly fading generation of heroes who fought for a better world for you and me. As James Holland puts it in his book *Heroes*:

> We can still find inspiration from their heroism. ... For the generation who were born from the embers of the First World War ... coming of age offered little cause for celebration. ... Mere boys found themselves facing life-threatening danger and the kind of responsibility few would be prepared to shoulder today. These people were an extraordinary generation. The majority of those who fought were not professionals, but civilians who either volunteered or were conscripted as part of a conflict that touched the lives of every person in every country

involved; ordinary, everyday people. ... When one said goodbye to friends or family, one might well be doing so for the last time, so worrying over trivialities was pointless, and petty disagreements and squabbles were cast aside. *Carpe diem* was not a catchphrase but a state of mind and this meant that most people were far better at living back then than we are today. As every year takes us further away from the war, so we become more selfish, more materialistic, more interested in the minutiae of life. ... The generation who fought in the war should not be forgotten, not now, while many still live, nor in the future when they are gone. We should learn from their lessons and remember and recognize the enormous sacrifices they made.

My gratitude goes to British author Roy Nesbit, and his friend and fellow writer and editor, Dudley Cowderoy, for their encouragement and their belief in the telling of my father's story. Roy Nesbit's detailed historical knowledge of the Second World War has been invaluable. I am also indebted to Roy, a busy author himself, for finding the time to review and comment on the manuscript. He generously supplied me with photographs from his private collection and permitted me to use the original tape that my Dad recorded for him for his own chapter on Dad's skyjack incident, 'The Flight of the Heron', in his book *Torpedo Airmen*. Both Roy and Dudley became great friends of Dad's, and they firmly believed the title of this book should be 'Larger than Life!'

My South African mainstay has been Professor Louis Changuion, long-time family friend and author of many books on South African history. I am thankful to him for taking time out of his busy and prolific writing schedule to read through, edit and comment on the manuscript; for providing me with valuable historical facts and source material; and most of all for his friendship and for spurring me on when my confidence was flagging.

My thanks go to Colonel Graham du Toit for his time taken in furnishing me with valuable SAAF archival research material for the appendix, and also to Mrs J. Williams at the Commonwealth War Graves Commission who, with a little sketchy information from me, very efficiently found and forwarded the exact details I required. Also, my thanks go to Paul Johnson and Hugh Alexander at National Archives for allowing me to use excerpts from RAF Official Records documents. Many thanks go to Florence Sheperd at HarperCollins Publishers for giving me licence to use their publication.

ACKNOWLEDGEMENTS

To author James Holland, my warmest thanks for his inspirational writing, and for so graciously giving me permission to use his words.

Bene grazie to my dear friend Mila Naccari for her professional translations of Riccardo Alessio's Italian letters. My gratitude also goes to my lifelong friend Sue Kuhn and my niece Lauren Van Heerde, for their efficient and resourceful help in helping me track down various people across the world for their permissions.

I am also grateful to my husband, Mike Morkel, for his frequent reassurance, for his patience, for being an intelligent sounding board, and for his tidy self having to live in a house strewn with scraps of paper and piles of books, memorabilia and photographs. Thanks to him and to my children, Abigail and Benedict, for living with my sudden blurted-out flashes of thought and obsessive ramblings, which seem to be in the territory of writing.

My thanks also go to Father 'Willie' Villabrod of the Catholic Church in Haenertsburg, and Frank Tilley, for giving me copies of and allowing me to use their memoriam and eulogy from Dad's funeral service.

Thanks to my brothers, Gary and Michael, and to my sister-in-law, Shirley, for giving me a corner in their homes when I needed a sanctuary in which to write.

I am indebted to all those at Pen & Sword who have assisted me in materializing this dream. My special thanks go to my publishing administrator Laura Hirst and my editor Linne Matthews for their incredible patience, enthusiasm and professionalism.

My sincere appreciation goes to my literary agents at Talk To Me, Monica Seeber and Sandile Ngidi, for accepting my manuscript, and especially to Monica for her patience with all my queries and inserts and her professional comments and editing. Thank you for giving my book the wings to fly.

I asked one of Dad's grandsons, Dean, to sum up his Papa's character in a few words. He spontaneously recited this beautiful piece:

Who would attempt to fly
with the tiny wings of a sparrow
when the mighty power of an eagle
has been given him?[1]

Gail Strever-Morkel
Uniondale, South Africa
October 2012

Prologue

The name 'Strever' means 'he who strives greatly to achieve that which he set out to do.'

Over a period of twenty years prior to Dad's death, I coerced him into sitting down and relating his life story to me. Every time I visited him at the country cottage on Stanford Farm near Haenertsburg, where he lived with my wonderful 'los-kop',[1] funny, long-suffering stepmother Andrée, I would insist on an hour or two of 'life-story time'. It was the perfect time and place. He was, as he put it, 'in the departure lounge', and his life had come full circle to the place of his childhood dreams, the magical Magoebaskloof. He playfully pretended to find my badgering irksome, and would protest on tape with, 'OK, you bloody slave-driver', or 'My daughter, who professes to love me, is pressurizing me out of my mind here.' Although I know he secretly quite enjoyed the attention, I also know it wasn't always easy for him, emotionally or physically. Over the years, as he grew old and tired, speaking for long periods made him breathless, and our sessions grew fewer and shorter. But his memory, especially of his early life, remained remarkably clear. He always used to say, 'As one grows older, the very early memories become clearer than the more recent ones.'

Dad had the knack of flavouring even the most mundane or difficult and trying events with humour, amazement, adventure and a big unforgettable laugh. Our sessions were often marked by much amusement and laughter, especially at night after he had had a few whiskies – although, this was not a good time to get him on a subject or a person he didn't like, and when, on the odd occasion it happened, Andrée (knowing what was coming) would get up and say, 'Oh God, here we go again. I'm going to bed.' Dad didn't suffer fools or 'poops' or 'drips' well. His most frequent sayings to us, his children, were: 'It won't hurt for long', 'There's no such word as can't', 'Where there's a will there's a way', 'Don't, be such a ninny, man', and 'Pull up your socks'.

My pressing need to record his story was motivated by two reasons. The most obvious was to retrieve it and save it from dying with him; the other

PROLOGUE

was because, until his later years, my father's war story was forbidden in any conversation, at home or in company. My poor mother, who was so proud of him, had hell to pay if she mentioned it or pulled out the newspaper cuttings and memorabilia. An example of this obsession of his was when the handsome and modest Lieutenant Colonel Don Tilley,[2] one of his long-standing Air Force friends, was being interviewed on Rhodesian television. Don, who had a distinguished wartime record himself, was attending a SAAFA (South African Air Force Association) conference in Bulawayo, and during the interview, which was being broadcast, he deflected the conversation away from himself and started talking about my father's skyjacking incident. My father was furious. He walked out of the room and would not speak to Don for years, although later this was something they both could laugh about.

My father, thankfully, mellowed with age. He loathed anything that smacked of boastfulness, self-promotion or vanity. To use wartime terminology, 'shooting a line' and 'glory seeking' were abhorrent to him. As James Holland writes, about the WWII veterans, 'They never thought of themselves as special because everyone had been in the same boat.' God forbid if, as children, we were seen to be showing off or boasting about something we had accomplished. It just wasn't done in those days; it wasn't honourable.

When I was growing up, what was referred to as 'his war story' was shrouded in mystery. I thought he went to the war to learn to be strong, so that he could fight off burglars and bogeymen to protect his family. To me, he was invincible; nothing could ever harm us when he was there because he had a 'war story'. There was always total safety in his presence – not only his large 6ft 2ins stature but also his full-blooded, ebullient, quick-witted character, expressed through a laugh that seemed to rise from the very centre of the earth; wholehearted and booming, contagious and unmistakable. He was always easy to find in a crowded hall! His hot temper, when sparked, was definitely something to avoid. A strong sense of justice, integrity and honour underwrote his life's creed.

This project of mine has been a long time coming. The recording sessions started in 1983 and ended with Dad's death in 1997, at which time, having obtained his dusty correspondence file from his Bushveld office, I managed to send numerous questionnaires to his friends and acquaintances for more information. The response was heart-warmingly amazing. Addressees who could not add much to my quest put me in contact with others who could. I had replies and help from places all over the world – Bermuda, Canada, England, Italy and New Zealand, to name a few. I also travelled from Cape Town to Harare to interview as many people as I could – especially those of Dad's generation before they passed on because, when I had asked Andrée if she had any names and contacts, she stated, in her blunt and funny way, 'No, man, they're all dead!'

So, armed with a lot of encouragement and helpful information, I felt ready to put this book together. But then life threw me a curved ball and my project had to be shelved. It is now ten years later and one night recently, lying awake at three in the morning, I realized that I might be lying on my deathbed saying, 'I never wrote the book,' feeling I had not honoured all those who had helped me with their time, effort and encouragement. The next morning I tried to find something on my computer from this early research, and to my surprise I found a file entitled 'Interviews'. Thinking that it was only a few pages, I pressed the print button and, to my astonishment, a hundred pages spewed out of the printer.

The story is primarily straight from the horse's mouth and mostly in Dad's words and voice, with some necessary interventions from me, as co-narrator.

Chapter 1

In at the Deep End

*Who strives always to the utmost,
for him there is salvation.*
Johann Wolfgang von Goethe, *Faust*

Bristol Beaufort Mark 1, in the air. Author's collection

> Daylight torpedo attacks against enemy shipping were some of the most hazardous operations ever undertaken by a pilot. To successfully launch a torpedo a pilot had to fly towards an enemy vessel at low level and into the teeth of fierce anti-aircraft fire.
>
> Gill Catton, 'The Legend of Don Tilley',
> article in *Wings Over Africa*, 1978

At last, someone on the ground responded; we were in contact.

'Am approaching with no landing gear ... badly shot up ... have ambulance ready ... injured man on board.' As I lined the aeroplane up with the runway at Luqa airbase on Malta, I shouted to my crew, 'Prepare for crash-landing!'

ON LAUGHTER-SILVERED WINGS

The ground seemed to rush towards us as I attempted to bring her in as gently as possible; I apprehensively held off … and held off … and … slam, we hit the tarmac with the most God Almighty noise. The metal underbelly of this heavy aeroplane screamed as it tore along the runway. Grit and gravel flew in at us like shrapnel. It seemed like an eternity before the deafening noise and bone-shaking motion stopped and the aircraft came to rest. The sudden stillness and silence fleetingly held us inert and incredulous before our adrenaline jolted us into reality and action again. We got the air gunner, Sergeant Bob Gray, who was wounded, out and into the ambulance. Then Pilot Officer Bill Dunsmore, the navigator, Sergeant Dick Ellis, the wireless operator, and I climbed out of the wreck that had somehow brought us back to safety. What we beheld was unbelievable. The aeroplane had 240 holes and gashes in it and the tyres were completely shot off of the wheels, leaving just a few strips of rubber. Thank goodness we hadn't had our undercarriage.[1] The aerials and struts on the wing for the ASV [Air Service Vessels] radar, a secretive piece of equipment, had been completely shot away. After assessing the damage an army tank was sent for and the plane was towed away to the scrap heap.

The wreck after the crash-landing at Luqa, Malta, 1942. Author's collection

IN AT THE DEEP END

Close-up of the wreck: note that the pilot's hatch has been shot away. Author's collection

But let's begin at the beginning ...

It all started after six months at an operation training unit in England, where I had recently celebrated my twenty-second birthday. My first posting was to the Far East. I was to fly from Portreath on the Cornish coast via Gibraltar and Malta en route to Ceylon. Leaving Portreath, we were three aircraft and crews in our formation. Leading was Pilot Officer Minster, followed by Pilot Officer McSharry and me. The seven and a half-hour flight from Portreath to Gibraltar was slow and laboured, because in those days we had no 'George' [a second pilot or automatic pilot]. It was an awfully long time to sit at the controls of a Beaufort – pilots sat and 'polled' the thing by hand all the way. All of us were also loaded to the hilt with aircraft spares, with the auxiliary tanks full to the last teaspoonful of fuel, and all our kit. We eventually landed at Gibraltar at three o'clock in the afternoon on 10 June 1942.

That evening we were given a substantial briefing and told: 'Due to the furious Jerry aerial activity around Malta you are not leaving in the morning tomorrow but at midday, so you will get into Malta at last light. You are to fly south from Cape Bon to avoid the Isle of Pantelleria, because there is a squadron of enemy Me109 fighters waiting there to attack the Allied ferry

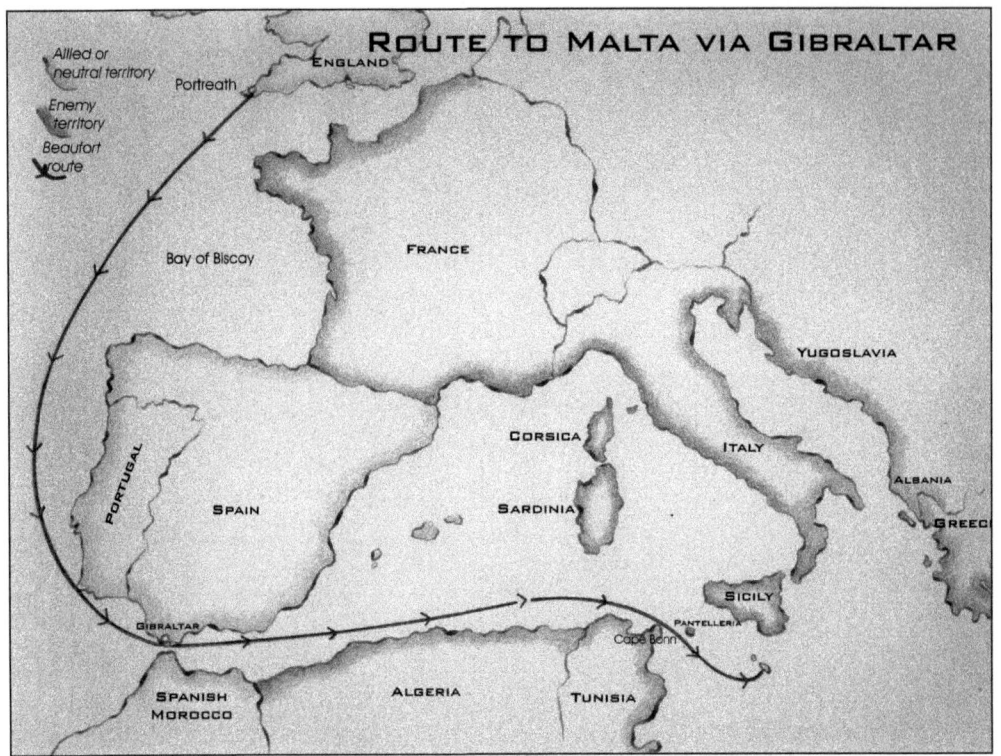

Route to Malta.

aircraft going to Malta. You are to keep a good lookout. Approximately an hour out of Gibraltar you will see a sight you will probably never see again. You are to maintain complete and utter radio silence, you are not to divert off course, and you are to proceed as planned.'

The next day we prepared to take off for Malta at midday, and midday in Gibraltar is hot – damned hot. Being such a rookie, straight out of training, I was not prepared for the effect that the heat has on air density and altitude; the hotter it gets, the longer it takes to lift an aircraft off the runway. However, at midday all three of us lined up on the runway, again with a full load. This runway lies across the narrow neck of Gibraltar from coast to coast, and our leader, Pilot Officer Minster, took off. I watched him and said to myself, 'What the bloody hell is he playing at?' He went off the end of the runway over the sea and was flying about 5 feet above the water weaving between the ships in the bay, and in astonishment I kept thinking, 'What is the idiot doing?'

Well, McSharry then took off and did exactly the same thing. Then I set off, not understanding what these blokes were doing until I was three-quarters of

IN AT THE DEEP END

the way down the runway. The bloody aeroplane was barely lifting, my wheels were still heavy on the runway, and this baby didn't want to fly. I heaved her off at the end of the runway and she just mushed through the air. I had to do exactly the same thing as the other pilots and weave my way gently between the ships until I eventually got up enough speed to climb to 1,000 feet.

We set course for Malta and about an hour and a half out of Gibraltar we saw the most spectacular sight below us. It was the June convoy (Operation *Harpoon*) from Gibraltar, sent for the relief of Malta. There were six merchantmen, twenty-six destroyers, with cruisers, one battleship, four corvettes, two minesweepers and two rescue ships. This huge nautical convoy, in formation, was a magnificent sight on the clear blue Mediterranean.

We continued on our way and came into Malta in the approaching darkness of evening. We landed at Luqa Aerodrome, and before our propellers had stopped, a torpedo on its trolley was being wheeled up to the aircraft. Mechanics with spanners were undoing the long-range auxiliary tanks, and there were shouts to open up the bomb doors. We hadn't even alighted. I said, 'My God, what's going on? We're supposed to be going to the Far East!'

A torpedo being loaded in a bomb bay, Malta. Courtesy of the Roy C. Nesbit collection

The reply was, 'No you're not, you're staying right here.' So our kit and all the spares we were carrying were offloaded. At our debriefing we were told, 'There is no way that you are going on to the Far East; you are to operate out of Malta. Your aircraft will be operational in the morning and you will be on standby for bombing raids on the Italian Fleet.'

A Beaufort in the parking pen, with Army personnel loading a torpedo, Luqa, Malta.
Courtesy of the Roy C. Nesbit collection

COPY OF OPERATIONAL RECORD BOOK: 217 SQUADRON, RAF STATION, LUQA, MALTA
The squadron first arrived at Luqa, Malta, today. The following crews composed the first detachment ...
More crews of the squadron arrived from Gibraltar today as follows:

IN AT THE DEEP END

2/Lt Strever, with P/O Dunsmore, Sgt Gray, Sgt Ellis.
P/O Minster, with Sgt King, Sgt Bowyer, Sgt Mogehonus.
Sgt Fenton, with Sgt Wallworth, Sgt Watson, Sgt Railton.
Sgt Carroll, with F/Sgt Parkes, Sgt Weaver, Sgt Wright.
P/O McSharry, with Sgt Augustinus, Sgt Brown, Sgt Wilkinson.
Sgt Smythe, with Sgt Dodd, Sgt Walls, Sgt Cornell.
The journey from Gibraltar was uneventful.[2]

Malta at this time was still beleaguered. It was a helluva long way from Allied territory.

It was one small lonely island in the middle of the Mediterranean. Only a hundred miles from the enemy in Sicily, but a thousand miles, on both sides, from the Allied forces at Gibraltar and Alexandria in Egypt. There was no food, no fuel and no spares to keep the war machinery running, so two convoys carrying the much-needed supplies were dispatched for its relief, one from Gibraltar in the north-west, and one from Alexandria in the south-east. This was done in an attempt to split up the enemy force should there be an attack.

Tragically, as we were to find out a few days later, it was an absolute disaster: they were both heavily attacked by the Italian fleet and only two merchantmen out of some forty ships from Gibraltar got through to Malta. The Alexandria convoy was forced to turn back to Egypt.[3]

On Sunday, 15 June, as the Italian fleet came out of harbour, we were woken up in the morning at 0200 hours, briefed, and sent out to look for our aeroplanes. And believe me; this wasn't as easy as it sounds because at this time Malta was pretty well disorganized.

For security reasons the aircraft were hidden up roadways in bombproof parking pens some miles from each other. These pens were walled on three sides by stone, which was built up to about 10 feet high. It was pitch dark when all nine pilots and crew piled into the bus. As we drove around dropping off crews, tractors were pulling aircraft out of the pens, which were all miles apart. We drove around and around the area three times and we were the last crew to find our aeroplane.

I was horrified to find it in a terrible mess. The windscreen was so dusty I couldn't see through it, so I refused to take off until it had been cleaned. This caused a delay. The other aircraft were already taking off, forming up and whizzing off, and we were still cleaning our bloody windscreen! We eventually get into the plane and I find that there are no brakes. Yet another delay while we get compressed air into the brakes. By this time we had lost ten to fifteen minutes and here was Strever and his crew, all alone!

ON LAUGHTER-SILVERED WINGS

> These take-offs and landings were marked by none of the ordered ritual of the modern aerodrome. There was no formal signal from a control tower, no stately departure or arrival in measured rotation.
>
> Ian Hay, *The Unconquered Isle: The Story of Malta*

However, I taxied out, took off and set course. When setting course I wasn't very confident about the accuracy of the readings I was getting because the mass of metal on the torpedo affects the magnetism of the compass. And there was not another aeroplane in sight.

We were supposed to attack the Italian fleet at first light by approaching from the dark side, where the enemy would be shown up against the light side of the sky. Well, we flew, and we flew, and we flew. It got lighter and lighter with still nothing in sight but the dark blue Mediterranean below. At our ETA [estimated time of arrival] there was still nothing in sight. So I decided to turn 90 degrees to port and flew for another ten minutes, still with nothing in sight. Then I returned to our original course, now flying back the way we came. After eight minutes we saw on the horizon what looked like little smudges on the sea.

Now, before leaving the training base at Leuchars in Scotland to take up my posting in the Far East [Ceylon] and having to fly via Malta, I had asked other crews who had returned from operations, 'What must we expect? What is really going on out there?' They reassured me with, 'Oh, don't worry, you new boys will be led gradually into things, you'll be taken out in formation to do a couple of anti-sub patrols along the enemy coast and a couple of mine-laying trips in the approaches of the enemy harbours, which you will only do when there is cloud cover. If any fighters jump you, you just pop up into the clouds and fly back.'

This had lulled me into a false sense of security and I don't think that it had penetrated my brain that we were about to encounter the heavily-armed battleships of the Italian fleet. So, as green as anything, flying at about 1,000 feet for better visibility, not down at 20 feet, where I should have been, we approached the 'smudges' on the sea. As more and more 'smudges' emerged they started looking alarmingly big.

We must have been a good 6 miles from them when I called out to my navigator, 'Bill, are you bloody sure there are none of our ships around here, because it seems as if those blokes are signalling to us?'

'No!' he shouts over the noise of the plane. 'They didn't brief us on that; there are none of our ships around.'

Soon I realized what was happening and shouted, 'By God, it's the bloody 10-inch guns of the battleships aimed and firing at us.'

IN AT THE DEEP END

There was now no mistaking their intention. All of a sudden we were flying into black smoke, with explosions all around us, but fortunately we couldn't hear them over the noise of the aircraft. In spite of not having the support and comfort of the other eight Beauforts around us to disperse the anti-aircraft fire, I decided that since we had found the fleet we would go in alone.

So I flew on.

There were two battleships escorted by two cruisers and seven or eight destroyers – eleven warships in all. The escorting ships were ahead of the battleships, forming an avenue on either side of them. I decided that the safest place to go was right down the centre of the avenue, straight towards the battleship, reasoning that if they fired at me they were going to hit their own ships on the opposite side, which would surely make them more cautious. Well, I can assure you they didn't desist.

I went in low, right between the escorting ships, head-on to the battleships, aimed – ascertaining from the bow waves that the battleship was in motion – and dropped my torpedo and shouted to Dunsmore in the front, 'Fire the Vickers gun straight ahead!' (My torpedo just missed the bow of the ship and I realized that, having already been attacked by our strike force ahead of me, the ship was stationary and had a bloody false bow wave painted on it to fool the enemy.)

The flak coming at us from all these ships was indescribable. There was tracer, green, orange, blue ... the water was boiling around us. There were big black shell bursts exploding. As I jumped the battleships from bow to stern I felt a crack in my back and ... shoomph! ... the hatch was shot off over my head, filling the cockpit with flying shrapnel and leaving gouges in the instrument panel. The turret at the back was shot away. A 20mm shell hit us on the wing root, where Bob Gray was manning the Vickers gun, and the shrapnel ripped into his thigh. Incredibly, the aircraft kept flying.

Phew! ... We got out of that lot! Now, you must realize that because of all the messing around looking for the fleet, we had only attacked at about 6.30 and it was now 6.40, about twenty-five to thirty minutes after sun-up – which was a very hazardous time to be flying around the Mediterranean. However, now that we were clear of them I assessed the damage. Shells had broken all the hydraulic pipes, there were holes all over the place, no radio, and we were in a terrible mess. The only things working were the engines and the control surfaces, the fuel lines, and the oil pipes, and there we were, stooging around an empty Mediterranean sky.

With no radio contact, a seriously defective aeroplane, fuel gauges close to the empty mark and a wounded crew member, I said to my navigator, 'Well, what do we do?'

He threw up his arms. 'I don't know!'

ON LAUGHTER-SILVERED WINGS

This wasn't a lot of help and made me lose confidence in him at that stage. I then attempted to fathom the correct course and ETA to Malta. I reasoned that on the original course I must have finished up north, or to port, of the target (the fleet), and that if I flew the reciprocal course I would end up to the left of Malta. So I turned 90 degrees again, and off I set.

Malta is a very small thing to find in the middle of the Mediterranean.

It was a devastating journey. All of us were suffering from the aftershock of the terrible shoot-up. I had gone into the attack in a sort of dream, not knowing what to expect. My reaction afterwards is something difficult to describe; I had knots in my stomach when I realized how close to death I was in those brief moments that I flew through those ships. I looked out, and there were one or two little cumulus clouds in the sky; it was a beautiful, clear sunny day over the Mediterranean. It all seemed a bit surreal. Then I saw a little dot on the sea and I whispered to myself in disbelief, 'No, is that Malta?' I was about 30 miles to the south of it and as I neared it I recognized that it was, in fact, Malta.

As I approached, with no long-range radio, I remembered the small emergency R/T set radio we had, which only operated within a circuit range of about a mile and had no distance range at all. So I got into the circuit range, anxious to make contact with someone on the ground. I then remembered that the hydraulic pipelines had all been shot away, so I had no undercarriage. After trying the emergency system, which didn't work, I knew we must prepare for a belly landing. But we could do nothing until I could hear a voice on the other end of my radio transmitter.

COPY OF OPERATIONAL RECORD BOOK: 217 SQUADRON, RAF STATION, LUQA, MALTA
15.6.42

Nine Beauforts took off from Luqa at 0415 hours this morning to attack the Italian battle fleet about 200 miles east of the island. The enemy fleet was in two sections, the leading section consisting of two battleships two cruisers, and seven destroyers.

W/C Davis led his formation of F/O Goodale and Sgt Hutcheson in to attack the first section, encountering intense flak but, it is believed, scoring two hits on a battleship. A piece of shrapnel hit F/O Goodale's aircraft, setting off a Verey cartridge, which was quickly extinguished by the navigator and W/OP.

Sgt Lynne attacked with Sgt Downe from the opposite direction. He scored a hit on the same battleship, while Sgt Downe hit a destroyer, which was last seen listing to port.

IN AT THE DEEP END

F/O Aldridge attacked the leading battleship in the southern force, or first section, scoring a hit.

Sgt Nolan also attacked alone, aiming his torpedo at the same battleship, but observed no results.

Lt Strever attacked the northern force, dropping amid intense flak, which wounded his gunner Sgt Gray. He had to belly-land on his return.

F/O Stevens attacked last. When he arrived, one battleship of the southern force was crippled, with a destroyer making a smoke screen around it. He attacked this ship but could see no results due to the smoke screen.

Chapter 2

Ignorance is Bliss?

Being in Bomber Command was not the most dangerous wing of the RAF – torpedo bombers, with a 17.5 per cent chance of surviving one tour took that honour – but the statistics are nonetheless astounding. Of the 125,000 men who passed through Bomber Command, about 55,000 were killed – a tenth of all British and Commonwealth war dead. Of course, most men had to believe that they would be spared, that death could not befall them, but the odds were stacked against survival, and witnessing the plane in front dissolve into a ball of fire and debris did little to ease the nerves. As Bishop says, 'The swing of the scythe was impressively arbitrary.'

James Holland, review of Patrick Bishop's
Bomber Boys: Fighting Back 1940-1945

I returned exhausted, with all my senses shaken from the first operation, and having been up since two o'clock, all I wanted to do was sleep. So after the debriefing I went to lie down. But try as I might, sleep eluded me. Every time I closed my eyes I could see streams of tracer, hear the explosions and feel the buffeting of the blasts. I lay on my bed and thought, 'Well, if this is what we are in for, we are not going to make it through this lot ... not with operations like that, and especially without a fighter escort to cover us.' My confidence of lasting this war out was seriously diminished. 'So much,' I thought, 'for being led gently into anti-sub operations.' I had gone into that first mission with no qualms; now I knew what to expect, and it wasn't at all encouraging.

I was now short of an aircraft and minus my gunner, Bob Gray, who had been wounded and was in hospital, but Bill Dunsmore, my navigator, was still with me. (Bob Gray eventually recuperated but never flew with me again.) In any event, on that same day, 15 June 1942, the other crews continued to go out on strikes, after which we all were given two days' leave to rest.

IGNORANCE IS BLISS?

COPY OF OPERATIONAL RECORD BOOK: 217 SQUADRON, RAF STATION, LUQA, MALTA
At 0930 hours F/O Minster and Sgt Fenton took off to attack two cruisers with destroyer and E-boat protection just south of Pantelleria. F/O Minster dropped too close, but Sgt Fenton, after his attack, saw columns of smoke issuing from the centre of one of the cruisers.

Nine Beauforts took off at 1930 hours to attack the remainder of the Italian Fleet. Unfortunately, the search proved fruitless, the squadron returning at 0100 hours (1400 hours). During the day, the squadron was joined by S/L Gibbs and a detachment of 39 Squadron from the Middle East.

16.6.42
Squadron is resting after yesterday's efforts, but standing by.

17.6.42
The Squadron was released for forty-eight hours. Aircraft are at readiness again.

The break gave me a chance to experience life around me on the island, and get used to the food rations: three slices of bread a day, one teaspoon of jam, no butter. Occasionally there was some goats' meat but we mainly ate tinned bully beef. A clerk would be sitting at a table just outside the mess, marking off your ration for the day. The cook in the officers' mess was quite magnificently inventive with the bully beef – he had at least fifteen to twenty different ways of serving it up.

It was midsummer, so extremely hot. Mosquitoes and sand flies were in abundance on the island, and were a bloody menace.

Of course, there were the incessant air raids, and the alerts were almost continuous. The anti-aircraft Bofors guns, which were dotted all around us, were blasting away as the enemy came over, so there was almost always a deafening noise surrounding us.

> 'A lull of any length,' said one of the gunners afterwards, 'produced a silence that was almost frightening.'
>
> Ian Hay, *The Unconquered Isle: the Story of Malta*

There were about four major raids every day, each lasting about two hours. The Jerry would come and bomb the hell out of the place, leaving bomb

ON LAUGHTER-SILVERED WINGS

The Army filling in holes in the runway at Luqa airfield. Courtesy of the Roy C. Nesbit collection

craters all over the runway, so the army troops would roar in, fill up the holes, and stamp them down in an attempt to keep the aerodrome operational. The aerodromes were, of course, primary targets, and the runways would be so potholed with craters that sometimes it was impossible to take off or land.

One night, the air raid warning sounded and down to the shelters we went (which we were obliged to do). The bombs were going off and we could hear and feel the thuds, crunches and explosions above us. While we were sitting down there waiting for the all-clear … all of a sudden, there was a big … whoomph … and thump … right at the entrance to the shelter, filling it with dust and muck. When the all-clear went out we emerged through this stuff and found a huge crater with a 500lb bomb stuck in the middle of it. Thank God, it hadn't gone off!

We returned to our billet, where three of us shared a room, stumbling along in the dark because of the blackout. The building had a flat Mediterranean style roof, and I went in first with a torch, thinking, 'Hell, there's a lot of dust around here.' I shone the torch onto the opposite wall and noticed that there were no mosquito nets over the beds, then I shone it down on our beds and there were heaps of bloody rocks on them. Then I shone it up to the ceiling and there was the nose of a bomb sticking through it. So we scarpered out of there very rapidly to the mess and told the wing commander there, who was slightly inebriated, 'There's a bloody bomb stuck in our roof and no ways are we sleeping there tonight.'

'My goodness,' he said in a restrained, pukka English way. 'Well now, we must get rid of that!' So off he toddled into the darkness.

IGNORANCE IS BLISS?

Bomb damage, Valetta, 1942. Courtesy of the Roy C. Nesbit collection

ON LAUGHTER-SILVERED WINGS

We decided that we couldn't let him go and fiddle around with it on his own, so off we went to help him. We climbed up on the roof and, sure enough, there was the back end of the bomb lodged in the roof. He says, 'No trouble, just grab hold, chaps.' He grabs the fins of this bomb and gives a big heave, we help him lever it out of the hole and he then starts rolling it off the roof, at which point I hurled myself to the other side of the roof, put my head down and waited for the big blast to go off and blow our barracks to smithereens.

But nothing happened.

I stood up and saw something, which looked like the top of a round cigarette tin, lying on the roof. I picked it up and asked, 'What's this?'

He said, 'Oh, I dare say, I don't know.' So I threw it back onto the roof.

Well, it was eleven o'clock at night by now and, with the bomb lying right next to the place, we definitely weren't going to sleep there, so we all went to the mess and slept on chairs and settees. At about three o'clock in the morning we were woken by a big crack and bang – and the next morning we found that the little round thing I had thrown down on the roof was in fact the detonator, and it alone had blown a hole right through the roof. Thank God we'd moved the bomb away from it.

The hole made by the bomb in the roof of the barracks. Author's collection

IGNORANCE IS BLISS?

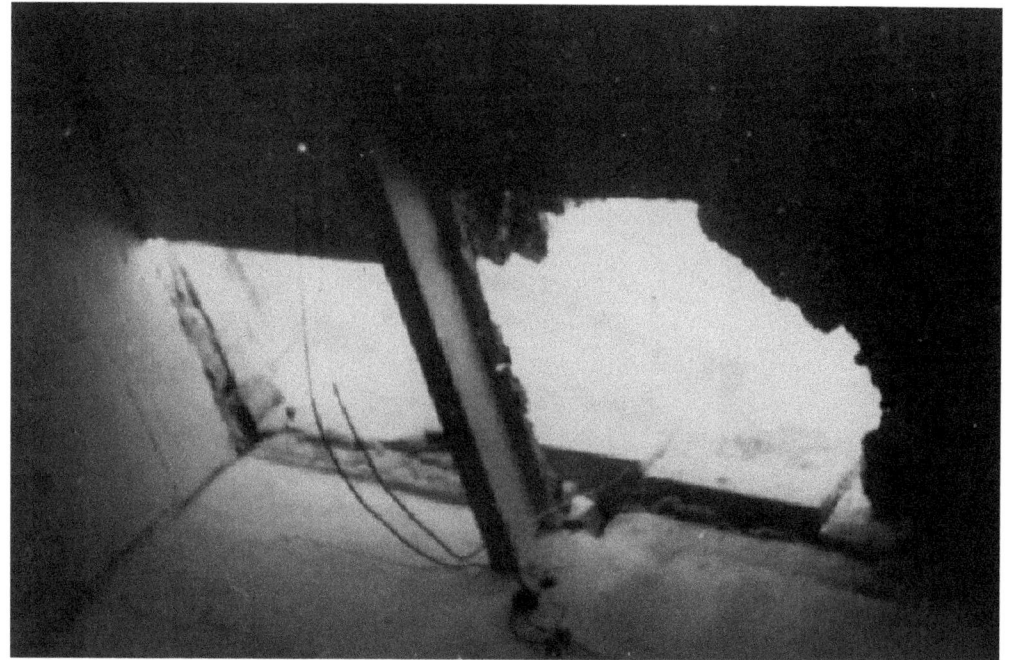

The hole made by the detonator in the roof of the barracks. Author's collection

Another time, a bomb had landed and penetrated the dome of the church at Musta, which, as the third largest suspended dome in the world, is the pride of Malta. At the time there was a full congregation and service in process; the bomb miraculously failed to explode. To this day, in Malta, divine intervention is firmly believed to have played its part in this episode.

On another occasion, feeling a bit soporific from the incredible heat of the afternoon, Bill Dunsmore and I were having our afternoon tea in the mess.

> The raiders continued their methodical visitations, coming over, as someone said, for breakfast, lunch and tea, and sometimes muscling in for supper.
> Ian Hay, *The Unconquered Isle: The Story of Malta*

Yet again the air raid siren sounded, and we, being a bit complacent, casually got up and sauntered towards the door. As we pulled the door's blackout curtain aside to go to the shelter, we got the shock of our lives. There, at about a thousand feet above us, was a whole formation of Me110s and Ju8s coming across the aerodrome, with their bombs already dropping out of them. Well, you've never seen two blokes move so fast in your life. We threw ourselves under a table in the mess, literally shaking and holding our breath, while this lot went over us.

Then, the next morning, while our Spitfires were sitting too high at 25,000 feet, over the enemy came again and did exactly the same thing – and they got away with it. However, that afternoon, when they returned yet again, the Spitfires were waiting for them at about 3,000 feet.

> The Hun ... possesses one commendable trait – he loves order and regularity. ... The Luftwaffe applied the same tidy and methodical routine to the bombing of Malta. They had certain and favourite targets – the dockland, the three cities, the island aerodromes, and Valetta itself – and these they attacked in mechanical rotation on an unvarying time schedule. So regular were some of these operations that in the course of time it became possible for the observers of the Flight Command, upon the approach of the raiders, to indicate which particular quarter of the island was about to be bombed. A red flag was hoisted over the threatened area, and its inhabitants took cover. The rest of the population could, and usually did, stay put and enact the role of spectators.
>
> Ian Hay, *The Unconquered Isle: The Story of Malta*

I have never seen such a sight in my life, watching as this formation of Luftwaffe medium bombers approached their target, the Spits peeling off and going in after them. We saw at least three or four enemy aircraft shot down as they were pulling out of their dive after trying to drop their bombs. Of course, there was a lot of shrapnel from the anti-aircraft flak that was going up from the ground at these planes above us, so we had got into the habit of wearing our tin hats to protect us from all the shrapnel that would descend to earth again.

There were the odd quiet times, when I would go with the boys for a swim in the beautiful clear Med water.

There was no transport, so we walked the 4 or 5 miles down to Sliema, or occasionally caught a lift downhill on a *gharry*, a cart pulled by horses.

Coming back uphill, the load of all of us was too much for the horses, so we walked back.

At this time, Field-Marshall Erwin Rommel, the commander of the German Afrika Korps, had got up to El Alamein on 25 June, and he had a long line of supplies coming across to him from Europe. It was thus our job to stop these

IGNORANCE IS BLISS?

Sailing and swimming on Malta: Ted (standing) and Dunsmore (seated), 1942.
Author's collection

ships from getting through to North Africa to feed, re-equip and resupply him. Now, remember, this was June/July '42, and we were the first striking force to come out from Malta and attack these Italian supply ships. Before we arrived, the supply convoys used to sail straight across the Med, but once we had hammered a few of them they started going around the toe of Italy, down the Yugoslav and Greek coast, hugging the mainland from about 5 miles offshore, and then across to North Africa.

They were not quite out of our reach, but we could only attack them when they were near the mainland, on only one section of the Greek coast, which

ON LAUGHTER-SILVERED WINGS

Ted in a gharry *on Malta, 1942.* Author's collection

was about 50 to 70 miles long. I must admit this close proximity to enemy coastline worried me a bit, because they could more easily have a fighter escort. Luckily, the squadron had already done two strikes in this region without encountering any fighter counter-attack. I mean, they were just 5 miles offshore, so why they didn't have fighters escorting them in this section I have absolutely no idea.

This brings me back to our vulnerability on the bombing missions. We [the squadron] had already had so many losses. On one of the operations/strikes, which I didn't go on, three out of eight crews didn't return; we lost a lot of good chaps.

> These men were forced to play Russian roulette as they sat in their cramped aircraft, risking death at any moment.
> James Holland, review of Patrick Bishop's
> *Bomber Boys: Fighting Back 1940-1945*

In fact, at the end of the six-week period (mid-June to the end of July), our losses were seventeen out of twenty-five crews that made up 217 Squadron and its replacements. All sorts of odd crews were coming in to assist us; this

IGNORANCE IS BLISS?

included 86 Squadron, and 39 Squadron from Egypt. It was not a happy situation. I will mention here that we had a sergeant pilot who had a fair amount of operational experience and had already been awarded the DFM [Distinguished Flying Medal], but on Malta, after a couple of these trips, he went LMF [Lack of Moral Fibre]. Of course, today his reaction would be more kindly called Post Traumatic Syndrome. He just refused to fly anymore and was promptly taken out and sent back to Britain.

We were eventually given a couple of Beaufighters to escort us on about our third or fourth operation. They were the night fighters on Malta stationed at the Ta' Qali airbase. The idea was that as we went into the attack they would attack the flak ships, or escort vessels, with their 20mm cannons, while we dropped our torpedoes. This, in fact, they did, and it was a most successful operation; we came out with hardly a scratch. However, the Beaufighter escort boys didn't like it because the flak was so severe, and there was a lot of animosity about it. That was the last we saw of them!

Six days after that first bombing operation, I was still without an aeroplane and two crew members. In any case, on 21 June, McSharry went out with the squadron on another operation, and a piece of shrapnel came through his perspex hatch and punctured the side of his face, severing a facial artery near his cheekbone. He had to fly an hour and three quarters back to Malta and the blood was pouring out of him. When he finally came in to land he was so weak and dizzy that he was at the brink of passing out. As he landed, the aircraft just bounced this way and that way, all over the place. We watched in trepidation as he finally brought the plane to a standstill, amazingly still on its wheels. We shot across to it, got McSharry out, as white as a sheet. He was very smartly given a blood transfusion and was sent to hospital. He had lost so much blood that it had soaked through his parachute and was actually dripping from the bottom of the aeroplane onto the tarmac. He told me afterwards, 'I couldn't hold off, all I could see was a shifting, swirling runway going up and down in waves.'

So with him out of action for a while there was a spare aeroplane and a crew without a pilot. I was now able to get flying again. I still had my original navigator, Bill Dunsmore, but was without my gunner, Bob Gray, and my wireless operator, Sergeant Dick Ellis, who had tragically been shot down on another operation. So Sergeant John Wilkinson, a wireless operator, and Sergeant Ray Brown, a gunner, both New Zealanders and both from McSharry's crew, joined us and we were back on operations.

We went out on our next operation and I got the feeling that our commander wasn't very keen on finding the target. We seemed to just stooge around over the sea looking for a merchantman, but we never found one and so returned to base. I wasn't very perturbed about this, and tell it with no fear of being thought of as a coward, because I thought, 'Well, that's two trips we've survived, even though on the second one we hadn't made a strike.' This gave me a bit more heart. The third trip we made was a piece of cake. We went into the attack and caught them completely by surprise. There was a merchantman escorted by two Italian frigates. We went in, dropped our torpedoes, made a hit on the ship, and were out again before they realized that they were being attacked.

Well, that was the third one we had got through. We did a couple more operations, in which we experienced quite a lot of flak, but they were nothing like my first mission through the Italian fleet, and my confidence was rising.

You asked me how it felt going out on a bombing operation. Well, that first operation, where I described going through the Italian fleet, was a classic example of 'ignorance is bliss'. I hadn't a bloody clue what I was going into, but I promise you, once you realized what war was really about, when ordered into action again your whole body reacts to the build-up of nervous tension. When called up for an operation, there was the preparation, the briefing, going out to the aircraft, getting into it, lining up on the runway, taking off, forming up into formation, and flying across; all these kept your mind busy but once you were on your way, which usually was for some length of time before sighting the target, you had time to think about what lay ahead, and your stomach would gradually knot up. You would have this very peculiar feeling. It's a feeling not unlike the kind of shock you feel after an accident or traumatic event. You would also have momentary thoughts of not really wanting to go on with the job and wishing there was a way out of it. On the other hand, you knew that you *must* go on with it, and ... it's very hard to describe ... all these feelings would sort of churn around your stomach. However, you only felt like this until the target was sighted; in that instant it's all that exists, it absorbs the full power of your concentration as you go straight in ... when you come out safely or still flying, and are able to get back to base, there is the most incredible feeling of euphoria. You just can't believe you've made it again.

Chapter 3

From Whence He Came

> A day unremembered is like a soul unborn, worse than if it had never been. ... So any bits of warm life preserved by the pen are trophies snatched from the dark, are branches of leaves fished out of the flood, are tiny arrests of mortality.
>
> Laurie Lee, *I Can't Stay Long*

'Mommy, Mommy, look how far I can widdle,' Teddy, my father, called in excitement.

Esther, his mother, was about to enter the Mafeking (now Mahikeng) chemist's shop. Mafeking was a dry and dusty little town in the far northwest of South Africa, near the border with Bechuanaland (now Botswana). Teddy was standing on the landing of a set of stairs, which led up from the street-level pavement to the shop entrance. This lofty position was just too tempting for the three-year-old – he had to test his growing prowess. When his mother turned around, he was on the edge, pants down, proudly aiming at the street below. Blushing with amused embarrassment at his brazen public display, his rather Victorian mother gave him his first lesson in social etiquette. The year was 1923.

Ted's mother, my grandmother, was born Pride Esther Woodward, the last of ten children, in Port Elizabeth, on 9 September 1895. Her name, Pride, was her mother's surname; she wasn't partial to it so everyone called her Esther. Her father, my great-grandfather, Alfred William Woodward, and her mother, my great-grandmother, Elizabeth Ann Pride, had met and married in England, where Alfred was working as a nurseryman at Kew Gardens in London. They had immigrated to South Africa, docking in Port Elizabeth sometime in the early 1890s.

Esther had seven sisters and two brothers. Of all her siblings it was Elizabeth Jane, the eldest, known as Lizzie, Katie, the third born, and the second-to-last born, May, who were closest to Esther. Of all the aunties, Lizzie, Katie and May had the greatest part to play in Teddy's life. Lizzie

Alfred Woodward (foreground, left) working in the nursery at Kew Gardens, mid-1880s. Author's collection

took on the role of his maternal grandmother and, as an only child for his first eight years, Teddy relished spending time in her home with his cousins. I never met my Great-Aunt Lizzie as she died before I was born. Dad, though, held loving memories of her, always recollected in nostalgic and grateful tones, with, 'Ah! I'll never forget how wonderful she was to me. I always had a place in her home.'

I did know Auntie Katie. She was small and stooped by a curvature of the spine, terribly forthright, and would scold anyone who neglected her with, 'I thought you were dead!'

In spite of her physical handicap she could walk anyone off their feet on a trip to town. She had married into the Baylis family of the Old Victoria Theatre in England, which apparently made her feel above everyone else, and when one of her nephews phoned to tell her he was getting married, she said, 'Is she a little English girl?' She wouldn't have accepted a Dutch or Afrikaans wife. She was definitely eccentric … as old age crept up on her and her memory faded, she would wander off to town on her own at all hours of the day and night and get lost. Once, she was found in the middle of the night wandering around Park Station.

Auntie May, like Katie, was diminutive but, unlike Katie, was the sweetest, dearest little old lady, with a tiny voice, sweet little laugh and delicate little movements. She married George Farrell, who was fittingly also of small stature (my parents nicknamed them 'Little Darby and Joan'). They were devoted to each other all their lives. Auntie Katie and Auntie May adored and worshipped my dad, 'Teddy', as they called him, and he, caring deeply for them, would not miss an opportunity to visit them as they grew older. When he visited them, his teasing would have them chuckling away, hands over mouths, with exclamations of, 'Oh, really, Teddy!' and 'Oh, no, Teddy!' They seemed of another time and a world long gone.

A few months after my grandmother's birth in 1895, the family moved from Port Elizabeth, for reasons unknown, to the famous diamond-mining town of Kimberley. After diamonds had been found there in 1871, all and sundry had flooded into the area seeking wealth and prosperity. The town, something akin to an American Wild West frontier town, had sprung up around them.

The Woodward family in Kimberley before the siege, 1895. Back row, from left: Katie, Lizzie and Fanny. Middle row, from left: Great-Granny Woodward with my grandmother, Esther, on her lap, then Lily, Harry and Doris. Front row, from left: May and Daisy. Author's collection

ON LAUGHTER-SILVERED WINGS

Four years later, the family were caught up in the 1899 siege of Kimberley, during the Anglo-Boer War. My granny was four years old during the siege, which lasted from October 1899 to February 1900.

During those four months, the drastic food shortage led to rationing, the need to slaughter horses for meat, and the erection of soup kitchens. There were outbreaks of scurvy. The mines were shut down and there was no work. Women and children were sheltered down the mines during the Boer shell bombardments by their famous Long Tom. I have wondered what my grandmother, at the age of four, made of all this – certainly, hardship was for most in those days a way of life, accepted and expected. She was of this generation, the thread of her formative years beaded with uncertainty, moving about, struggle and loss, and she grew to be quietly long-suffering and adaptable, qualities that would prove indispensable in her life with my grandfather.

The family survived the siege but within a year of its end, their mother died. Aunt Lizzie, the eldest daughter, then became the maternal figure in my grandmother's life and, it seems, in everyone else's. After his wife's death, my great-grandfather moved the family from Kimberley to Johannesburg and set up his own nursery at Natalspruit, near Germiston.

Here my grandmother Esther grew up under the care of Lizzie and her other sisters. She met and married my grandfather, William (Bill) Henry Strever (known to us, his grandchildren, as Pop Strever), before the First World War.

They had come from different worlds, she from a large family of women and he from a family of men. She was slimly built and delicate, with Victorian manners and a gentle, retiring disposition. Having become an excellent seamstress she was always well turned out. She had fine-boned, chiselled features and deep-set eyes, which often reflected a lurking sadness. She was also, by all accounts, lovingly accepting of her husband's restless and impulsive nature. He was a big man with a solidly-built frame, strong handsome features, and a square, dimpled chin. He looked out intensely from hooded eyes, a Strever hallmark. (This somewhat strident look was inherited by my dad, who used it to very good effect – he jokingly called himself 'eagle-eyed Eddie'.)

Unlike Esther's very English family, Pop Strever had come from German stock. His father, my great-grandfather, was one of two Strever brothers who had come out to South Africa sometime in the mid-1800s. One was Harold William and the other was my nameless great-grandfather. There are three

FROM WHENCE HE CAME

Some of the Woodward family, Johannesburg, 1908. Back row from left: Alfred Woodward, Harry and husband of Fanny (name unknown). Middle row, seated from left: Lizzie, Fanny holding son Edwin's hand, Katie with daughter, baby Lillian, on her lap, and May. Foreground, from left: Doris and Esther. Author's collection

quite different accounts of how and why they came to South Africa. In the first, they fought in the Crimean War of 1853-1856 as mercenaries, and were rewarded by England with land in South Africa near Stutterheim, in the Eastern Cape. The second account has them as deserters who fled the Crimean War and their country, arriving in Durban. The third has them quite ordinarily arriving with the 1820 settlers and settling in Queenstown. However, Great-Grandfather Strever apparently died in the early 1900s, and the truth lies buried with him. I know which version my dad would have preferred to believe!

After my Great-Grandpa Strever's death, Great-Grandma Strever married 'Pop' Higgins, became Great-Grandma Higgins and had three more children: Archy, Frank and Lulu. To my dad she was Granny Higgins and her last three children, very confusingly, were his step-uncles and aunt. Although they were a little older than Dad, their closeness in age made them more like cousins. They lived on a large dairy farm called Hatherley, near Bronkhorstspruit, north-east of Pretoria. Hatherley had belonged to the early entrepreneur Sammy Marks, who had set up the first factory in South Africa there. The site on which the factory was built became known as 'Eerste Fabrieken'. Dad remembered it as a glassworks factory, and when visiting his grandparents as a child he loved collecting the discarded lumps of coloured molten glass that lay scattered in its surrounds. Times spent at Hatherley were the happiest of his childhood.

The farmhouse and its world provided room for boyhood dreams and adventure. As an only child for most of his early years, Dad gained a sense of family from visits to Hatherley, and of all his memories these were imbued with a radiant warm glow, and left indelible impressions on all his senses. It was rich with company, chatter and laughter: the women bustling around the large kitchen table; the mouth-watering smell of the Sunday roast sizzling in the old wood stove; and the intriguing stories Pop Higgins told at mealtimes.

One such story was about an old prospector who was a friend of Pop Higgins. Dad says, 'Well, this poor old chap, while prospecting near

Granny and Grandpa Higgins at Hatherley, 1925; Ted on Granny Higgins' lap, his young step-aunts and uncles behind them. Author's collection

FROM WHENCE HE CAME

Granny and Grandpa Higgins on Hatherley, 1925. Author's collection

Hatherley, had found a huge piece of glassy rock in late 1933. He was convinced he had found a massive diamond, and had told Pop Higgins that he had found a fortune in a diamond and described the stone to him. The stone very soon vanished – lost or stolen – and he went quite demented. Soon after this, the news broke of a very large diamond that had been found, the now famous Jonker Diamond.[1] Could it have been the same stone that the old prospector had lost? As it was not 'dug up', or unearthed, but was lying on the surface of the ground, this is not impossible. Sadly, this twist of fate, Dad insists, 'brought the old prospector to insanity – Pop Higgins told us that he went quite mad after this incident.'

Outside in the back yard, about 50 metres away from the house, was the long drop toilet, which meant 'you couldn't hold out too long, otherwise you had to run like hell to get there in time.' Away from the farmhouse there was only the hot silent stillness of the Transvaal winter days, where the gentle cooing of doves and mooing of cows wafted on the air. Tall gum trees cast long cool shadows over the garden and all was wrapped in the sweet-sour smell of dairy cows. Beyond the garden, the Bushveld was to be explored, and promised unknown treasures. Dad remembers his excitement at finding some old Boer War brass shell cases on one of his hunts. Here, too, adventurous games of cowboys and Indians were played with his cousins.

When Pop Higgins retired from the railways where he worked, he opened a sweets and cool-drink kiosk at the Eerste Fabricken station. Dad loved being with his Pop and eagerly helped out in the little shop at the station whenever they visited. When he was thirteen, Dad's family visits to Hatherley ended because in 1933 they moved to Natal. However, in 1939, when Pop Higgins died, Dad returned with his parents to Hatherley to see Granny Higgins and the family. It was to be for the last time because shortly after this, war broke out, the calling of which changed his life forever.

My grandfather, 'Pop' Strever, Dad's father, came from a large family of railway workers. (The South African Railways, being the main source of long-distance public and goods transport in those days, was a big industry.) He was born in 1889 in Aliwal North, in the Eastern Cape on the Orange River, and had four brothers and one sister: Gustave, Charlie, Edward, Albert and Gertrude. Gustave worked all his life as a cleaner on the railways and was always described, sympathetically, as 'Poor, old Gustave, you know he wasn't all there, he was a bit simple.' My mother recalled that, 'When Pop Strever was dying at our house in Bulawayo in 1949, poor Gustave arrived

Teddy in his cowboy outfit picking flowers at Hatherley, with Eerste Fabrieke behind him. Author's collection

on our doorstep one night quite unexpectedly. He stood there looking lost and forlorn, with a little tin of *padkos* (food for the road) in hand. He was the only brother to come and say goodbye to Pop, and had used his discounted staff train ticket to come up all the way from South Africa.'

About Charlie, a ganger on the railways, my father talks with amusement: 'Poor old Uncle Charlie, he was terribly browbeaten by his wife, Aunt Lizzie, the fiery redhead. He had to ask her for money for cigarettes because she insisted he hand over all his pay to her. When she died,' he continues, now laughing, 'they found a fortune of £5 and £10 notes hidden all over the house, especially in the linen cupboard, where she had stuck them between the sheets, pillowcases and table linen. She was quite a character!'

Edward, known as Uncle Ted, became the station master in Queenstown in the Eastern Cape, and then in Oudtshoorn, famous then for its ostrich feather industry. He was one of the brothers to whom my grandfather Pop remained close, and after whom my dad is named. He gave my mother a huge white ostrich feather fan on ivory stays as a wedding present; it is now one of my family treasures.

I don't know much about Albert except that he was the manager of the Pretoria railway workshops. His son, Dad's cousin Bertie, an idealist, was, at the time of the Second World War, a member of the 'Brown Shirts' or Stormjaers (storm troopers) of the Ossewabrandwag (oxwagon guard), and spoke fluent German. It was the time of the infamous (or famous, depending on where one's allegiance lay) Robey Leibbrandt, a South African Nazi who stood vehemently opposed to General Smuts' support of Britain. So did the storm troopers of the Ossewabrandwag when war was declared in September 1939. Apparently, Bertie wasn't the only anti-British member of the Strever family when war broke out; it had both the country and the Strever family divided. This caused a big rift between the brothers who supported the Allies and those who supported Hitler and refused to fight for England.

This became a dark secret that hung over the pro-war side of the family and was only ever spoken about in hushed tones, which shrouded it in mystery and gave it a thrilling sense of foreboding for us as children. The Nazi sympathizers remain nameless but I suppose they can be identified by their absence from Allied duty. This family chasm has left us grasping at fragments of our history. However, perhaps because his brother Bobby joined the Air Force, Cousin Bertie did, thankfully, eventually enlist as a medic – that is why he is mentioned by name, having been absolved by denouncing his fascist leanings.

Esther, c.1910.
Author's collection

Pop, c.1906-1907, in early British uniform, from the Bambatha Rebellion period.
Author's collection

FROM WHENCE HE CAME

All I know of Auntie Gertrude is that she married a Thomas Bell, who was nicknamed 'Ting-a-ling' by the family.

But to go back a bit: when Pop Strever met my granny just before the First World War he had by this time already served in one war. At the age of seventeen, in 1906, he served as a saddle-hand in the Bambatha Rebellion in Natal.

Pop's medal for his service in this war is missing from the rest of his 'gongs', but the distinctive magenta and black ribbon on his medal bar remains as proof of his effort. Pop also fought in the First World War as Private W.H. Strever of the 12th Infantry. He served with the Pretoria Regiment in German South-West Africa and then with the artillery in German East Africa, Egypt, Palestine, France and Belgium. He survived, but returned home with lung damage and breathing problems from being gassed in the trenches.

10th Party Maxim gun course, Union Defence Force, School of Musketry, Tempe, November 1913. Pop is fifth from right, back row. Author's collection

ON LAUGHTER-SILVERED WINGS

Pop in the First World War. Author's collection

On returning home, Pop trained and qualified as a coachbuilder and trimmer (upholsterer) in the South African Railway workshops in Pretoria. He met and married my grandmother, Esther Woodward, and moved to Mafeking to work in the railway coachbuilding workshop.

By 1919, Esther was pregnant. She had already suffered a couple of miscarriages and the heartbreak of a stillborn son, named Edwin, whose lifeless little body was put into a mustard bath in a desperate but vain attempt to stimulate his circulation and revive him. This time, taking no risks, she insisted on going from Mafeking to Primrose, Germiston, near Johannesburg, where Auntie Phillips, a midwife, would help deliver her baby. Auntie Phillips, a relative of Pop's, lived on a farm in Primrose. There, in her old farmhouse, which stood on what were then vast open grasslands, my dad was born on 12 February 1920. With her slim build and delicate disposition, he must have been more than she had bargained for, emerging like a fat cherub weighing 12 healthy pounds. The successful delivery seemed a miracle.

He was named Edward Theodore, but all his life he was affectionately called Teddy by his family, and Ted by everyone else. Baby Teddy was Esther's pride and joy, and a series of subsequent miscarriages meant that he was her only child for eight years. After the birth, he was duly, and proudly, taken home to Mafeking, where he lived for the first three years of his life. The only memory he had of his time in Mafeking was the trip to the chemist shop, where he had quite unwittingly mimicked the famous bronze statuette *Mannekenpis*![2]

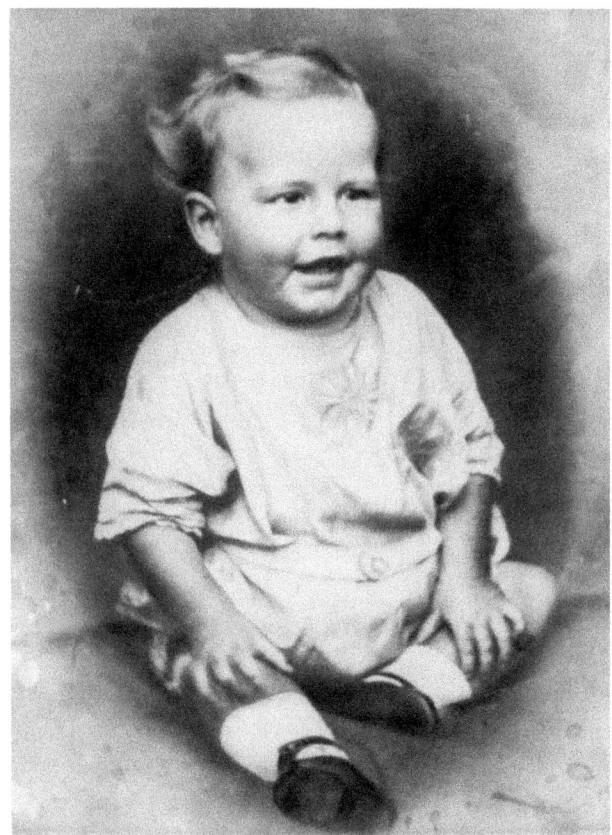

Teddy at seven months old, Mafeking, 1920.
Author's collection

Chapter 4

A Shifting Permanence

In 1923, when Dad was three years old, the family moved from Mafeking to Pretoria, where Pop Strever had decided to go it alone and start his own upholstery business. It coincides with the poor economic climate in the country at the time; 1920, the year Dad was born, was marked by the start of a serious economic depression that resulted in unemployment and wage cuts, and ended in 1923. It seems likely that this was the reason Pop Strever risked leaving what must have been a stable and secure job with the railways to start his own business in Pretoria. The move was made, a workshop was set up, and Pop got to work, measuring, cutting and upholstering.

From this time on it was Dad's father, Pop Strever, who was the prevailing influence in his unfolding world. He explains: 'He was the dominant feature in my life, which doesn't mean I wasn't close to my mother. She was the quiet, gentle, loving and caring anchor of my life. But Dad was the driving force in the family, and poor Mom had an awful lot to put up with in her married life. Although he was a laughable, likable, lovable man, he was generous to a fault, which made him a hopeless businessman. His open-handedness with money, when we desperately needed it, did not make life easy for my mother. He would pick up stragglers and stray people and invite them home when we had very little ourselves. He was also very impulsive, and it was nothing for him to come roaring in on a Saturday from work and say, "C'mon, let's go to Jo'burg," and boom, boom, boom, we were all piled into the car and off we went. Well, Mom put up with this unsettled lifestyle and I think she sometimes enjoyed these spontaneous deviations from routine life.'

At the age of four, Dad spent time watching his father at work in his upholstery workshop. When Pop cut out the fabric or leather pattern pieces with his big upholstery scissors, he would chew his tongue in a synchronized motion with each squeeze of the scissors. The imprint of this idiosyncrasy stuck with my dad, and chewing his tongue became mandatory for any bout of concentration – especially when he was about to take off or land an aeroplane (here the tongue chewing was accompanied with a wriggling in his seat, as if he was also lining himself up to the runway). Once, Dad was watching Pop cutting, and became very frightened when all of a sudden Pop shot backwards across the room, his body jerking violently. He had inadvertently cut through an electrical wire that was somehow lying under the fabric. Luckily, he survived. Little Teddy itched to also do things in the workshop, but his four-year-old attempts usually led only to grief and tears. One day, alone in the workshop, he decided to help his father, so he got a brace and bit and made a big hole right through the middle of one of the newly upholstered chairs. When his father came in he said, proudly, 'Look, Dad, I'm helping you!' – and promptly got a clout. He got the same reaction from his father when, with a mighty kick that would have stood Bennie Osler[1] proud, he sent a Lyle's Golden Syrup tin straight through the glass window.

But these shared days with his father in the workshop were short-lived. The upholstery business didn't succeed. As my father affectionately puts it, 'My Dad, as good a bloke as he was, was never a businessman, and the business was a disaster.' So in 1925 they moved to Johannesburg, where Pop had taken a job as a trimmer and painter with Connock Motors. At Connock's, his work entailed re-upholstering roof canopies and car seats, and repainting the bodies of the touring cars they had in those days. Dad states proudly: 'I think he was virtually the first person to do spray painting on the Reef.' Laughing, he adds, 'And he swore that from the first day of inhaling the toxic spray paint fumes, he was cured of his lung problem, caused by the gassing in the trenches in the war.'

While working at Connock Motors, Pop bought a small stand in the open countryside at a place called Townsview, which is now in the southern suburbs of Johannesburg. Here, he built a simple little dwelling for his family, with a cow dung floor. One day, five-year-old Dad, playing alone outside with matches, set the veld alight, for which, yet again, he says he got the 'hammering of my life'. He remembers going to his first school in Townsview: 'It was a convent called La Rochelle, and I was never happy there. All I remember of it is the little bottle of milk I took to school; it was like an Eno's Fruit Salts bottle with a cork in the top. I used to sit by myself in the corner, hating every bloody minute of it until school finished and I could go home again.'

It is here, in Johannesburg, that Dad remembers his maternal grandfather, Alfred Woodward, as 'a white-haired old man' who was staying with one of his daughters, Dad's Auntie May. He seemed to have no permanent abode, so moved around from one daughter to the other. Lizzie's daughter, my father's cousin, Edith, remembers when Grandpa Woodward had lived with them. She recalls the surprise dance parties they used to have, of which he obviously did not approve. He would storm out of his bedroom in his pyjamas and shout, 'You are all going to dance yourselves into hell!'

'He was a redhead,' Edith states matter-of-factly. When he died in 1926, at the age of seventy, my father, then aged six, remembers the terrible sadness of his mother and all his aunts. Alfred is buried in an unmarked grave in Brixton Cemetery.

Sometime during their stay in Townsview there was a lot of unhappiness in his parents' marriage, which might have been sparked by being so near to all of my grandmother's family. My granny left Pop and moved into a hotel. My dad says, 'I can remember staying in a hotel with her and she was very sad and upset. I had nightmares about losing her and being all alone. My obsessive thought was, what would I do if she died? So I decided that if she died I would hide her under the bed and never let anyone take her away from me. It's funny,' he continues, 'how these emotionally traumatic experiences affect you, and stick in the memory. Anyway, about ten days later they made up, but before she moved back home she took me to Durban. I can remember getting a blazer from my dad and into the pocket he had slipped half a crown, which was especially for me as pocket money. My mom took me to the little kiosk on the beachfront and I bought a little sailing yacht and had a lot of fun with it in the paddling pool. On our return they were reconciled and everything seemed fine again.'

Most of Dad's childhood memories of his father have three elements: a motor car, a mishap and making a plan – all of them recounted with much chuckling and laughter. 'My dad always had a motor car,' he recalls. 'One of the first cars he had was an old Renault, one of the first self-starter motor cars, which meant you didn't have to get out and crank it up to get it started. It looked like the 1905 models we see today, with the sloping front and open sides. When I was five I remember going with Mom and Dad in this old Renault to visit our friends, the Kingstons, who lived on the Van Rijn Estate mine near Benoni. It was a very bouncy ride – in those days, the roads were very rough. Main Reef Road was just a track through the bush that curved around the mine dumps. I'll never forget, as we came around one of the mine dumps the old car stalled right on the railway line, and heading straight towards us was a train from the other side of the mine dump with its whistle going madly. Dad shoved the car into gear and started it, and with a jerk, jerk, jerk, jerk we managed to get over the line just in time. My mother swore that she could hear the engine driver cursing at us as he whistled past.'

After about a year in Townsview, Pop Strever accepted a very good job offer in Pietersburg, in the then Northern Transvaal. He felt it would be a good thing to move away from all the meddling women from my grandmother's family because there was friction between the sisters, and some refused to talk to others. So Pop sold the little place at Townsview quite easily, for about £200, an enormous sum in 1925-26. It was duly paid into the trust account of the lawyer, a Mr Chivers, who promptly disappeared with it. Dad says, sadly, 'It was a big and bitter blow for Mom and Dad.'

Chapter 5

Like Father, Like Son

> Life is either a daring adventure or nothing.
> Helen Keller, from *The Open Door*

'Bill, dear, we're not going all the way to Pietersburg with *Molly* making this terrible noise, are we?' asked Granny Strever.

'*Ja*, man, it's fine, she'll get us there,' Pop Strever replied, quite unperturbed.

The car was firing on three cylinders and missing wildly, soon after setting off from Johannesburg to Pietersburg. They had got within 6 miles of Pretoria when *Molly's* engine blew up. My granny thought this was a most serious catastrophe – here they were, setting out on yet another move, with yet again an old car, with their few belongings and no money, thanks to Mr Chivers the *skelm* lawyer.

Molly was a 1918 Buick that Pop had bought while he was with Connock's. My granny had named her *Molly* and she loved the car. *Molly* had a canvas fold-back top, and was only a two-seater. My father loved this because he travelled perched high on the folds of canvas roof just behind the seats. With the wind in his face, eyes closed and outstretched arms, he was flying over the expansive Bushveld rushing past him. But this dreaming only lasted while *Molly* was actually moving, which seems to have been not all that often. The family had set off on this particular Saturday, *Molly* happily bumping along through the gum trees that lined the old Johannesburg-Pretoria road, until disaster struck.

'Now,' Dad affectionately relates, 'Pop was, to put it mildly, quite a harum-scarum sort of bloke but he was a fantastic character. Nothing ever got him down. You can't believe nowadays the trials and tribulations of motoring in those days. The old split rims that had a puncture every 10 miles – and there was no such thing as changing the wheel. It had to be taken off, the tyre patched up and pumped up again, and, oh, the nuts would get cross-threaded, it was really something in those days.'

LIKE FATHER, LIKE SON

What a load! Molly, *1925: Ted, second from left on Esther's lap and Pop standing behind them.* Author's collection

So, on this occasion, no trouble to Pop, he jumped out, took the sump off, removed the broken part (which happened to be a con rod), put it all back together and set off again … but on three cylinders, which had *Molly* spluttering and pup-pup-pupping. To my granny's dismay, Pop didn't stop in Pretoria but carried straight on past it, fully intending to go all the way to Pietersburg in this condition. Dad says, 'Incredibly, *Molly*, spluttering and missing, kept going on those bumpy, potholed dirt roads.' But she continued to have puncture after puncture.

That night the family slept out on the side of the road somewhere on the Pietersburg side of Warmbaths. The next day, Sunday, they continued in fits and starts on their way, having again puncture after puncture. At one stage, Pop gave some chap, who had helped him fix a puncture, a lift. He was going to Nylstroom, and had to sit on the back of the car. 'In those days,' Dad explains, 'you had to whizz down the hills to help you get up the other side.'

On approaching Nylstroom, Pop apparently did just this, and with the poor chap at the back hanging on for dear life, he shot right through the town before he realized it. The hiker had to walk back to Nylstroom. By late afternoon they had got as far as Ysterberg, a mountain between Potgietersrus and Pietersburg with two peaks. (These twin peaks, which create a 'saddle' between them, are believed by some to be 'Sheba's Breasts' from Rider Haggard's *King Solomon's Mines*.) They had travelled the last 5 miles on tyres stuffed with grass, having run out of tyre patches, tubes and everything else. 'This was another gimmick,' Dad tells me, laughing. 'In those make-a-plan days, in the event of no patches, the tyre was pumped up by shoving grass into it with a tyre lever, and getting as far as you could on it. As you can imagine, it wasn't a smooth ride!'

That Sunday in 1926, the family, and dear old *Molly*, came to the end of their tether at Ysterberg. Her old engine was still going on three cylinders but there were no more tyres. All this, and my granny still in silk stockings, dainty button-up shoes and fine frock ... so, with twinkling lights beckoning in the darkness from Pietersburg, so near and yet so far, they slept again in the veld that night. 'I'll never forget that place,' says Dad.

For the rest of his life he never passed that spot without reliving this childhood memory, and I never pass it without thinking of him, camping out there, aged five, with his mom and dad.

The next morning, they sent a message with a passing motorist into Pietersburg, to Pop Strever's prospective employer, Mr Parucatti (Dad's pronunciation: 'Paragetti'). Mr Parucatti owned the Pietersburg Garage, the Fiat Agency and the General Motors Agency, and promptly arrived to help them in his Fiat.

'Well,' exclaims Dad, 'I thought this Fiat was the last word in luxury. With all its knobs and buttons, and little red and green lights on the dashboard, I thought it was too wonderful.'

At last, after two days on the road, they finally arrived in Pietersburg on Monday afternoon.

Pietersburg was to be their home for the next eight years, the longest time the family stayed in any one place. At first they lived in a boarding house, and Dad was sent to kindergarten and then to Pietersburg Government School. During these school years his mother gave birth to his sister, Sheila Rose, born as tiny and delicate as Teddy had arrived big and sturdy.

At this time, Dad's world was rapidly expanding. His after-school activities included Cubs, and later, Scouts. Among his close childhood friends who shared these boyhood days was Wally Levy.

When, in 1998, Wally was reminiscing to me about Dad and the early Pietersburg days he said: 'My earliest memories of him were when we were in the Cubs in 1928. He was eight and I was six. We cycled to Cubs together because we lived near each other. I remember how excited we were when our Cub mistress, Mrs Du Plessis, took the Cubs on a train trip to visit the Johannesburg Zoo. Later, when we were in the Scouts, our Scoutmaster, Bernie Bulchers, used to take us camping to the experimental farm, out on the way to Witkoppies. We had great fun camping there two or three days at a time next to a stream that ran through the farm. Ted was always on those camps with us, he was very keen you know, and as I say, he was very friendly, outgoing and always a leader. You know, we were all from poor homes and had to make our own amusements. We used to go fishing at the bridge on old Marshall Street, and to save a tickey[1] we swam at the municipal dam instead of the public swimming pool. We

Boy Scouts, c1930. Teddy, top row, far right. Author's collection

Teddy with baby Sheila, Pietersburg, 1929. Author's collection

also played plenty of pranks; we used to sneak into the convent grounds and raid the peach trees.'

My dad, laughing, recalls, 'The first time we stole peaches we stuffed them down the front of our shirts, and all went home itching like bloody crazy!'

By standard three he was doing well thanks to a kind teacher, Miss De Waal. Standard four, alas, was to bring grief in the form of an 'old dragon', Zadie O'Shea, who, as Louis Changuion wrote in his book *Pietersburg; Die Eerste Eeu*, conducted her lessons in High Dutch. 'Whatever I did,' he insists, 'was wrong, and the more she hammered me, the more stroppy I got and the less I did; it was a vicious circle. My marks dropped, and she sent me to the headmaster's office so many times for cuts[2] that eventually Mom and Dad decided this was no good at all.'

So Dad was moved to The College of the Little Flower, a Catholic school run by brothers. This made a tremendous difference to his life. 'The brothers were fantastic. Brother Edward was especially good to me and from very low academic achievements in standard four, I moved up, with his help and

College of the Little Flower, Pietersburg, 1932. Teddy is front row, third from left, with Brother Edward standing at the back. Author's collection

encouragement, to second or third place in the class. Dan Collins, from Haenertsburg, always came first. He was a very clever chap, and no one could ever catch up to him. So the third and second position was always between a chap called Louis Hyams and myself.'

This was a happy time for Dad, but once again, it didn't last long. At some stage during these school years in Pietersburg, Pop Strever had set out, yet again, on his own in business. It was registered as 'W.H. Strever, Body Building, Trimming, Painting and Upholstering' and was situated in Maré Street.

Dad says, 'Once again, it was disastrous!' The decision of Pop's to go it on his own fell again in the wake of an economic depression, this one following the Wall Street Crash of 1929. 'It's so sad really, when I look back,' Dad says, 'because he was a very skilled craftsman and was very good at his job. He had got the reupholstering of the old canvas car canopies to a fine art and could do it in record time – it was normally a five-hour job at the factory. It entailed taking the canopy off, removing all the tacks from the bows, cutting out and stitching up the new one, inserting the pads, and then refitting it onto the car. Pop, with the help of one assistant, could accomplish it in one hour thirty-five minutes, from start to finish. But he just couldn't make a go of a business. He was a soft touch for anybody. If he had money in his pocket to buy material for a job and some bloke came and said, "Ah, Bill, I'm really pushed, have you got a tenner to lend me?" out it would come without any further thought about the materials he needed. Of course, he usually never saw the money again.'

So, by the end of standard five there was no money again. Aware that his parents were struggling to pay the college school fees, my dad volunteered to go back to the government school, and started high school at the dual medium Pietersburg Hoërskool, where, with the devoted help of his teacher, Mr Leo Saloman, who spent hours giving him free tuition after school, he apparently scraped through his standard six Junior Certificate. In 1933, in standard seven, he had a history teacher by the name of Mr D.J. De Kok, nicknamed 'Old Kokkie'. Although a 'delightful character', he would ramble on, and on, and on in his lessons, and the boys would groan when they had him for the last period of the day because he would be spouting forth long after the bell had gone. So Dad and his classmates got up to all sorts of shenanigans. Laughing, he says: 'We would pack our bags, and vie for a desk at the window. Then, as soon as his back was turned, we would chuck the bag out of the window, put red ink on a handkerchief, clap it to our nose and run up and say, "Sir, my nose is bleeding." He'd say, "Go on then, hurry up." We'd go, ducking under the window outside, pick up our bags and bugger off.'

Dad's weekends and holidays in Pietersburg were happily spent going on Boy Scout camps, picnics with his family and cycling with friends. His mother had insisted on his having the complete Scout's uniform and it took her four months to save the one pound five shillings to buy one. These Scout camps and cycling trips were enjoyed, 60 kilometres east of Pietersburg towards Tzaneen, in and around Haenertsburg and Magoebaskloof, a lush, misty, mountainous area historically steeped in local magic and folklore, the beauty and mystery of which beguiled and bewitched many a traveller.

It was in the majestic mountains of the Magoebaskloof that Chief Magoeba, in attempting to protect his claim to the land, was killed and beheaded by, as Louis Changuion informed me, a contingency of Swazi warriors who fought on the Boer side in the war of 1898. The area is also a place that caught the imagination of author-adventurers Rider Haggard and John Buchan, and which inspired much of their writing. It was called 'The Valley of the Mists' by Harry Klein in his eponymous book about his travels in, and love for, the area.

Camp at sawmill, Magoebaskloof, 1932. Esther is far left. Seated from left: friend, Sheila and Teddy. Author's collection

LIKE FATHER, LIKE SON

One camp Dad remembers was on the farm of the mining magnate Sir Lionel Phillips, to whom John Buchan had dedicated his novel *Prester John*. Part of the dedication reads thus:

> *To Lionel Phillips ...*
> *So take this medley of ways and wars*
> *As the gift of a friend a fellow-lover*
> *Of the fairest country under the stars.*
> Googoo Thompson, from *Between Woodbush and Wolkberg*

At another time, they camped next to a stream on Major Wolf's farm at Cheerio Halt. One day when they were swimming, lightning struck the water upriver. Dad describes it: 'The bolt hit me in the water, as though someone had clobbered me with a pole behind my knees. You've never seen blokes move so quickly ... talk about walking and running on water!' These camps were full of fun and adventure for him. The beauty of the place embedded itself in his adolescent soul, giving him an enduring love for it all his life. And thus began his lifelong dream to return and settle there one day.

Meanwhile, having caught the gold bug, Pop had got himself involved with two gold mines; one was the Ennis Mine, 40 miles out of Pietersburg, the other at Mampa's Kloof, in the Strydpoort Mountains. The old motor car *Molly*, after being fixed up in Pietersburg, was still part of the family in spite of her many more mechanical failures. Once, on the way to the Ennis Mine, north-east of Pietersburg in the Mooketsi area, Pop got lost in the bush, with the family, in old *Molly*. 'Because,' Dad explains, 'he was very obstinate and stubborn. When he was set on going in a chosen direction, he was going in that bloody direction and nothing would persuade him otherwise. He certainly wouldn't consider turning back to find the right road; he would just press on. This particular time, the road we were on trickled into a track and the track dissolved into a footpath, which had us eventually driving through the bush.'

In an attempt to find a road, Pop drove *Molly* on a cattle path, down a steep rocky incline onto a riverbed, and couldn't get her up and out the other side. In the excessive effort, the crown wheel in the differential broke a couple of teeth. So another night was spent stranded in the bush. For my grandmother it was sleepless, with the sound of lions roaring nearby.

Next morning, help came from a kraal settlement nearby. For a princely sum, the local chief got the oxen rounded up and harnessed to *Molly*. His oxen then pulled her for 3 miles to the Ennis Mine. Here, they were stuck for quite a few days without any communication with the outside world and had to buy chickens from the chief to eat as they had no supplies. (After this experience, his mother referred to a chicken as a '365' – one for every day of the year. 'We're having a 365 tonight,' she would say.)

Dad continues to describe in detail how Pop made a plan: 'Among the old mine things was an old forge that had been left there, a few bits of iron and an old engine. Pop, having borrowed some tools from the local chief, got to work. He heated up the crown wheel, took it out, drilled two holes in it, got some old valve stems, hammered them through, riveted them over, heated them up and fashioned them into the shape of a tooth. He then fitted it back into *Molly*, started her up, and it made the most incredible noise. This was not an engineered or precision fit, but my God it worked and got 35 miles to the nearest telephone, where it packed up again.

'At another time, she threw a bearing, a big end bearing, on a long trip to somewhere. Pop, got out shouting, "Bugger, blast and dammit," whipped the sump off, took off the broken bearing, cut a piece off of a leather strap, fitted it into the bearing, tightened it up and away we went. It remained there, in perfect working order for a good eighteen months. Pop's outlandish mechanical improvisations became legend.'

In those unregulated days of motoring, Pop had taught Dad to drive when he was nine years old, so that he could drive his mother to the shops, which he did frequently (she never did learn to drive). On one occasion, Pop, who was working at the business, urgently needed to get a message out to the chap on the Ennis Mine because telephones were non-existent in those remote areas. He decided that Dad, now thirteen years old, was capable of driving the 40 miles in the car on his own. So, unbeknown to his mother, Dad proudly and excitedly set out, with his friend George Myburgh, his cousin Stella and his five-year-old sister Sheila in the car with him. He tells the story: 'I was quite proud of this ... off we went on those dreadful roads; in those days the main road was a track through the bush. Well, we eventually got to the place, delivered the message and set out on our way back.'

He continues: 'Now, I could drive, but no ways had I ever driven that far before or had the responsibility of it. ... Fortunately in those days there was very little traffic. When it got dark I put the headlights on, but by then I was very tired, I had absolutely had it! I decided to get George, who had never driven before, behind the wheel to just steer, and I did the gears. Well,' laughing, 'it was hilarious; we were going all over the bloody road until he got the steering taped, when we managed to continue uneventfully. We arrived home at nine o'clock that night, and my mother was furious with Pop. That was quite an adventure and one I wouldn't have missed because it gave me a bit of confidence in myself.'

During this time Pop had both the mine properties checked by a geologist for deposits and had been assured they were good. He eventually managed to get a mining company interested in investing in the Ennis Mine when, as only Pop's luck would have it, he became desperately ill. He was rushed to hospital, where they found he had malaria, complicated by blackwater fever, and very soon he developed pneumonia. He was in a critical condition.

'At one stage,' Dad recollects, 'they said, he's gone, there's nothing more we can do for him, and according to Mom, they actually covered Pop up.'

Determined not to give up on him, my grandmother said, 'No, I can't believe it!' and begged the doctor, 'You've got to keep trying.' So Dr Taylor-Smith[3] worked for hours and hours on Pop. Suddenly, there was a flutter from Pop's heart and he was back again, albeit to remain a very sick man for some time, wracked with fever, frequent convulsions, vomiting and pain. Dad describes his father at that time as being 'an absolute, tottering, physical wreck'. This tragic turn of events meant that for a long time he couldn't work at all, and the family was broke again. So Pop, in desperation, was forced to sell his mining claims for a pittance, and Dad, at fourteen, had to leave school. This is where my father's childhood ends.

The doctor advised that the family move to the coast, but they had no money to pay for the train fare. Fortunately, my grandmother's sister, Auntie Katie, was able to supply the necessary funds for them to move, yet again, this time to the south coast of Natal.

Chapter 6

Turning Point

Before leaving Pietersburg for the coast, my grandmother sold everything they owned and, with Teddy's help, took Sheila, who was six, and a very weak Pop, on the train. They settled in Port Shepstone, a coastal village about 70 miles south of Durban. They arrived with nothing but a few clothes and a future filled with uncertainty. Auntie Gertie, Pop's sister, had found the family a small furnished cottage to live in, costing one pound five shillings a month. Here, while Pop convalesced, they eked out a precarious existence surviving on the few pounds they received from Auntie Gertie, Auntie Katie and Auntie May.

When Pop had mustered up enough strength to work again, he managed to rent a small paint shed on the side of a garage in which to set up a business. Owing to Pop's diminished stamina and resources, for the following two years, Teddy had to help his father in the paint shop. During those two years, in spite of being deprived of schooling and the opportunity to make new school friends, there were some happy times. Pop made some furniture, and Granny was able to create some semblance of a home around her, first in a flat they rented and then in a delightful little hillside cottage that was festooned with climbing roses and had a beautiful view of the Umzimkulu River flowing into the sea.

While Dad worked at the paint shop he did manage to make three friends: the local magistrate's daughter, Audrey, who was about his age, and two young chaps who worked at the bank. The four of them knocked around together in a little gang. During one of the school holidays, to the boys' delight, three of Audrey's girl cousins from the Northern Transvaal came to visit for a long seaside holiday. One of the boys, Jimmy, had a little two-seater car, and with a newfound freedom they would all pile into it and whizz off with the wind in their faces. They felt that the world was their playground. These carefree adolescent explorations towards new horizons

TURNING POINT

Pop with Ted, Sheila and friend on the beach at Port Shepstone, 1934. Author's collection

were spent picnicking in the countryside or having seaside fun on the beach in Margate, a few miles down the coast. For Dad, two life-changing events transpired from this happy holiday visit.

One was his first bittersweet taste of romantic enchantment. The object of this primary passion was Estelle, the youngest of the cousins, who had my dad, at the end of the holiday, mooning and pining for his lost love. It proved to be a pivotal influence. 'She said, "Ted, what are you doing with your life? You can't go on helping your father in the paint shop, you've got a good brain; you must finish your schooling and then study. You have the ability." It was her encouragement that made me realize I was wasting my life.'

Until that time, what he was doing was never questioned. Pop apparently had his own ideas about education because he had seen university graduates having to work on the roads during the Depression – he thought it was 'for the birds'. He was a craftsman and worked with his hands. 'Why should I push my son to get a matric, to end up working on the roads?' Thus unwittingly he was denying Dad the vital parental encouragement and motivation he needed in those important teenage years. Providence happily intervened, helped along by the encouragement and support of his mother.

By the time my dad was sixteen, Pop's health and strength had been restored, but in the business it was the same old story. Pop was too easy-going, gave too much credit, did jobs too cheaply and, as Dad sadly puts it, 'We never made the grade.' So Pop had to leave Port Shepstone to take a job in Durban. The family moved into a furnished room in a crummy place on Brickhill Road, opposite the Bakers biscuit factory. At times of hunger – and there were many – the sweet, warm, vanilla smell of the baking biscuits had Dad and his little sister Sheila longing and drooling for them, and they were the frequent subject of their dreams.

By this time, my granny was determined to help Dad find a job. So they watched the newspaper for any vacant situations. After six weeks of job hunting, he managed to find a very unlikely job for a 'rough and tumble boys' boy' at a salary of four pounds a month. It was as an assistant window dresser with the large, posh departmental store Ansteys in the town centre. According to Dad: 'My boss, Anton Gerlach, was the chief window dresser, an incredibly patient chap, who tried in vain to appeal to some artistic flair in me.' For the first two months, Dad's job was to work the letterpress in the basement. 'There were hundreds of cards to be made for sales. I had to set the press, place shiny adhesive paper onto cardboard and put it through the press, which would then die-cut the letters and emboss them onto the board. Well, this wasn't too bad. However, fetching and carrying corsets, brassieres and other bits of lingerie for display in the windows had me, at sixteen, scarlet-faced with embarrassment.' On Friday nights, all the shops would be open from seven to nine o'clock, and Dad enjoyed working on these nights. Town was crowded, the atmosphere was festive and lively, and people would make this a weekly evening outing. They all had fun after work and it was something to look forward to during the week.

In spite of the financial difficulties, his mother was adamant that he must 'get something behind him', so as soon as he had started working at Ansteys, in May 1936, she helped him enrol for standard eight at night school at the

TURNING POINT

Durban Technical College. He would go straight from work to school, finish there at nine o'clock and, to save the fourpenny bus fare, would walk home, which took him an hour. He says, 'I saved those pennies and eventually had four shillings and sixpence to put into my building society book. I felt terribly wealthy.'

At this time, Pop managed to get a more permanent position with a firm of motor car bodybuilders in Johannesburg. And for the first time, Pop, Granny and Sheila moved away without Teddy.

Soon after this, Dad managed to leave his unsuitable occupation at Ansteys and get a junior audit clerk position with an accounting firm, for seven pounds a month. For the first few months his job entailed adding up huge columns of figures in a book (there were no adding machines or calculators in those days), and this experience left him with an enduring ability to add up figures. After his expenses he had ten shillings to spend, which allowed for a once-a-month Saturday afternoon matinee at the Criterion bioscope down on the Esplanade, and on other weekends a tuppeny bus ride to the beach.

Ted in Umbilo Babies' baseball outfit, 1937. Author's collection

The boarding house eventually closed down, so he moved into the family home of his friend, Clarrie Lamble. Here he became part of the family. Clarrie was a keen cricketer and had played in the Natal Cricket Trials, and also played baseball for the Umbilo Giants with big names like Dudley Norse, a renowned Springbok cricketer who played baseball out of cricket season. Through Clarrie, Dad started playing baseball. He got himself an outfit, was given a second-hand glove, and played for the second team, the Umbilo Babies.

By this time he had sat his standard eight exams and had passed all the subjects except bookkeeping. He had done well in history, geography and mathematics but had failed bookkeeping by a few marks. The principal of the college, Mr Sutherland, who knew he was struggling to afford his studies, said to him, 'It is such a pity, Ted; you missed a scholarship by a few marks. I wish I could have given them to you.'

ON LAUGHTER-SILVERED WINGS

After a year on his own in Durban, Dad was missing his family, who were now all settled in Johannesburg. So he managed, with Clarrie's help, to get a job with a chartered accountancy firm, at eight pounds a month, in Johannesburg. Pop and Granny were living in a flat in Braamfontein, near the old Johannesburg Hospital. Between their flat and the centre of Johannesburg lay the old Wanderers Club playing fields (this area has all now become Park Station, Johannesburg's main railway station, but in those days it was very open and green).

In Durban, Dad had worked in a large firm and had been a little fish in a big pond (hence all the adding up) but the Johannesburg firm was small, and allowed him the opportunity to gain more advanced experience in accounting. He handled the switchboard and was given some of the bookkeeping to do. He continued at night school at the Technical College in Johannesburg, and re-sat bookkeeping and got a distinction. Having finally got his standard eight, he was able to apply to the School of Accountancy to do his CIS (Chartered Institute of Secretaries) studies.

In the meantime, Pop had, through his brother Albert, gone back to working for the railways and was sent to work in Bloemfontein. Dad had to move into the YMCA in Jeppe for a time, and then rented a room from his cousin Edith and her husband Lionel in Bezuidenhout Valley. To save the tram fare to work he bought a bicycle for seven pounds and paid it off at a pound a month, and rode to work. Lionel and Edith were very good tennis players and tried to teach Dad to play the game. However, this was a game at which he definitely never excelled.

At the end of 1937, the firm moved Dad temporarily to their Klerksdorp branch. They had a liquidation to do, and too much work. He took his bicycle and moved to Klerksdorp. He got on well with his boss, an elderly chap, and was soon asked if he would consider staying on permanently. When Dad explained that he would like to but had his CIS studies to do in Johannesburg, his boss then said: 'What if I got the partners to agree, and I offered you articles to become a chartered accountant and we will still pay you your salary? If you are prepared to move down here I will give you twelve pounds ten shillings a month, and get you articled.' This offer was like manna from heaven; he couldn't refuse, because young people wanting to be articled with a firm normally had to pay the firm to have them and they worked as a student with no salary. Delighted, and realizing what a gift this offer was, Dad responded excitedly, 'Yes!', like a shot.

The partners agreed and from the beginning of 1938, Dad stayed on in

TURNING POINT

Klerksdorp. To be articled for the five-year accountancy apprenticeship with a firm, the minimum qualification was either a matric certificate or the preliminary chartered accountant certificate. Not having a matric qualification he was advised to study for the preliminary chartered accountancy exam, on which he duly embarked. He says: 'With the promise of articles and assured of a living salary if I kept passing my chartered accountancy exams during my five-year apprenticeship, I could eventually become a professional chartered accountant. I now could see a future and was fired with enthusiasm.'

He stayed in a boarding house run by an elderly spinster by the name of Olie Lowman. This woman, a disabled polio victim, instilled in my father an everlasting admiration and respect for handicapped people. He describes her as 'the most fantastic woman'. (He always believed that there was an extra special dimension to the character of handicapped people.) A firm friendship grew between Dad and this 'fantastic woman' who, he insists, was a sort of mentor and a very good guiding influence in his life.

Ted in Klerksdorp, 1938 – after a hard day's work. Author's collection

He now worked very hard at night on his studies, often swotting until two in the morning. After work he cycled in and around Klerksdorp with a friend from the boarding house. They sometimes rode for miles out of town, as far as the Vaal River.

He also started to play rugby after work. In July 1938, at eighteen, he was called up by the Active Citizen Force for an army training camp, due in September. This was a terrible blow as he was in the throes of swotting for an exam to be written in November. He did everything in his power to delay enlistment. He appealed to the army, saying, 'Please, I don't want to get out of this but I've got these exams, they are most important to me and I don't want to disrupt my studies at this stage. Please can I have an exemption?' The answer was: 'No! Off you go!'

Klerksdorp only had a regiment of army infantry, the Regiment Delarey, named after the legendary Boer War General Koos de la Rey. In September

Family group in Klerksdorp, 1938. Author's collection

1938, he went to camp, with the vain hope of squeezing in some swotting time. The new recruits were housed in very large noisy dormitories and their days consisted of tough physical drilling, cross-country running, carrying rifles and packs and, as he puts it, 'By God, I got fit and with the few months of rugby training and then this camp I've never been so fit in my life!' As it was all in Afrikaans, Dad learned the language very quickly; most of the

TURNING POINT

boys were Afrikaans and he got on very well with them. He had a lifelong respect for Afrikaans-speaking South Africans, and often felt more at home with them than he did with the English. In spite of not being able to swot for his imminent exam, he thoroughly enjoyed the active, physical, outdoor nature of the army training. Within a month he was promoted to lance corporal. After the camp, he went back to work, sat his exam in November, and failed it. He was bitterly disappointed.

The following year, Dad again attempted to continue with his studies but he was sent by his employers to help out in their Johannesburg office and was again called up in September for an army camp. At the beginning of September 1939, the Second World War broke out. Dad, by now in his second army camp, had been promoted to company sergeant major of C Company of Regiment Delarey. He laughs, saying, 'I think it was this foghorn voice of mine that was my greatest asset as a sergeant major.'

Promotion meant that he was required to stay on longer for a sergeant major's course. However, he sat the preliminary CA exam again in

Ted with his mother after he joined the South African Air Force, 1940. Author's collection

November, and he remembers: 'This time I was very despondent because my studies had been interrupted again. I remember battling through the maths paper to about three-quarters of the way through, and couldn't fully complete one of the problems, left the exam thinking to myself, I'm never going to make it, and was convinced I had failed again!'

Believing strongly that he had failed his exam and thus didn't have a future in accounting, he decided to volunteer for the war effort. His army CO, a Captain Hartley, had volunteered and transferred from the Army to the Air Force in an administrative capacity. Although Dad had enjoyed his army experience, like many young boys his dream was to fly, so he went to see Captain Hartley and said, 'Sir, I want to fly. I want to volunteer and join the Air Force.' In the meantime, Pop had left the railways and started his own workshop, which wasn't going at all well. When Dad told his parents he was going to volunteer, Pop decided that since business was so bad, he would join up with his son. 'Well,' Dad says, 'I don't know how Captain Hartley worked it, but in January 1940, Pop and I joined the Air Force.'

Pop Strever in his South African Air Force uniform, 1941. Author's collection

Chapter 7

Bucking the System

In his naivety at twenty, my Dad thought if he joined the Air Force he would automatically become a pilot and fly. After all, what was the Air Force for but for flying aeroplanes? To his dismay he was enlisted to give orders, which he did for his first six months in the Air Force.

When he and Pop had joined up, Pop was sent to the aircraft workshops and Dad, having been in the regimental section of the Army, was put into the disciplinary training section of the Air Force. He was to be a drill instructor to the new recruits who, when they weren't at lectures, were on the parade ground. All day he was required to pound and square-bash them around the parade ground, do squad drill and physical training. He says: 'Well, being on my feet all day kept me fit, but I really hated it, so I kept badgering for a chance to apply for pilot training, but there seemed to be no way out of it.'

He was soon transferred to the Armaments and Aircraft (AMA) Depot, a few miles away at Robert's Heights (later known as Voortrekkerhoogtre and today Tswane Heights, or Thaba Tshwene). There, too, 'I was a bloody drill instructor!' he exclaims, 'drilling these blokes all day and taking them for regiment. It was ghastly! So I persisted in pressing for a chance to fly.'

Eventually he got his chance, and went before the selection board for pupil pilot training in the South African Air Force (SAAF). Fourteen hopefuls went in front of the board and three were chosen. He was one of the three (the SAAF was being so selective because they had a shortage of training aircraft). He says boyishly: 'I was so thrilled about this!' Then he learned that he had, in fact, passed his preliminary chartered accountant's exam, which he had been convinced he had failed. However, although it was now too late to sign up for articles, having the qualification probably helped in his acceptance for pilot training.

He went through the tough medical examination, which required a high

59

level of fitness, and passed it. Then soon after this, he received an unexpected note that shattered his hopes. The note read: 'Regretfully, you cannot go on to the Pupil Pilot Course.' He was at a complete loss at this rejection, as he had already satisfied all the necessary requirements, and he says: 'I don't know what it was, but I think the reason, quite frankly, was my family's lack of social standing; they were not from the upper crust or the top drawer.' Although life as he was growing up had been a struggle and a challenge, it had been spent in blissful ignorance of the influence or sway of social class. He had, of course, been aware that there were other people who were better off than he was, but as a youngster it had not impinged on his sense of self. Not until he was twenty, and in the Air Force training camp, did he become aware of that thing called 'class'. Although being in camp was a great leveller, and all the trainees were treated as equal and became buddies, they were a mixed bag. Now, Dad, aware that most of the trainee pilots were university students and came from privileged backgrounds, seemed to feel the inadequacy of his background. Many years later, at a South African Air Force Association Congress, he asked Colonel Bernard Thorpe, one of the elderly members, what the reason could have been. He assured Dad that it was due to the fact that he was under twenty-one. But my dad – being my dad – never believed it and always detested this thing called 'class'.

At the time of this untimely rejection, the CO said, 'Look, you can go into the Air Gunner Training Unit.' For Dad, this at least offered the opportunity to be airborne, the promise of some action and a way out of shouting orders. So he transferred to Course 7 of the Air Gunner Training Unit, in June 1940, and was sent from the AMA Depot at Roberts Heights back to CAFTD (Central Air Force Training Depot). 'Where I used to bash the blokes around the square, now I was being bashed around the square, but it didn't matter, we were going to get somewhere!' Many firm friendships were formed in these training days. One such friend was Bobby Rogers, who reminisced in a letter to me:

> Most of us felt like fish out of water on reporting to camp but your dad, being an army instructor, helped us in many ways to find our feet. We were also not a very well disciplined bunch on the whole. We were far more interested in fighting the Germans than performing on the parade ground. I think your dad was rather worried about this attitude to start with, being an instructor and all that, but he soon joined in. We found when

BUCKING THE SYSTEM

we reported to CAFTD that there were just not enough uniforms for all of us, so we started off drilling in 'civvies'. Then we were issued with the bare essentials, such as boots and greatcoats, and gradually acquired other items of uniform. So, one way and another, we were never the smartest chaps on parade – with the one exception! Your dad was by far the smartest soldier on Course 7 (especially compared to us civilians), he was tall, well built, had a great sense of humour and always ready to get up to mischief. He became a loyal, dependable and caring friend.[1]

From CAFTD they were sent to Ground School at Tempe Camp, in Bloemfontein. The place, according to Carey Heydenrych, in *Heck! What a Life!*, was 'large and bleak, hot and dusty and devoid of soul'. Here they started training all over again! Air gunner trainees were lectured on bomb aiming and air-gunnery theory. In Course 7, Dad said, 'I realized I was not alone in my quest to become a pilot, because all of us were aspirant pilots. I met and got to know some wonderful chaps who became long-standing friends; amongst them were Bobby Rogers, Don Tilley, Gill Catton, Wally Morgan, Joe Morgan, Cliff Evans and Hugh Sheldon.' The chaps were soon to find their lowly status was reflected in their 'thrown together' attire and decided to take matters into their own hands. Bobby Rogers[2] explains:

> At CAFTD we were all still finding our feet but by the time we got to Bloemfontein (Tempe), things became a bit more organized, or disorganized, as the case might be. In camp we wore whatever we could put together. We shared Tempe camp with Pupil Pilot courses. There was, of course, a wire fence between our two bungalow areas to separate the candidate officers from us plebs. The first thing we noticed was the smart barathea or gabardine uniforms worn by the pupil pilots and we demanded to have something similar. This request was turned down and it was explained to us that they were officers' uniforms whereas we were mere air gunners under training. To cut a long story short, we arranged for John Scott, a well known Pretoria tailor, to design us a uniform that differed only in that it had five buttons down the front instead of the usual four. We wore this uniform when stepping out until we qualified as pilots in Rhodesia months later.

ON LAUGHTER-SILVERED WINGS

Of the lack of discipline and the many humorous incidents, Bobby Rogers notes:

> I remember a chap Van Rooyen, built much like your father, who having been AWOL for the night, had returned, having climbed under or over the perimeter fence just before morning roll call parade. He was magnificently attired in white tie and tails, and still slightly under the weather. The bugle sounded just as he got into camp and, as there was no time to change, he went on parade in his tails. The officer in charge, one Peter Croft by name, marched up and down the ranks inspecting the motley crew on parade and eventually came to Van Rooyen, standing rigidly at attention and gazing straight ahead. He stopped and slowly looked Van up and down, goggle-eyed.
>
> 'What do you think you're playing at?' he eventually demanded.
>
> 'Who, me, Sir?' answered van Rooyen.
>
> 'Yes, youuuuyou! ...' he managed to get out through clenched teeth. By this time he was literally foaming at the mouth. Restraining ourselves from an outburst of raucous laughter was all that the rest of us could do to keep some semblance of order. Needless to say, Van was confined to barracks for a lengthy period. Not that it made much difference.
>
> I'm not sure whether instructors were dumb or whether they deliberately let us get away with murder – perhaps both. Trainees regularly slipped roll calls, knowing that the rest would cover for them. I remember at roll call parades, if members were absent we would fill the front rank and leave gaps in the rear ranks. We would answer 'present' when the names were called and cheerfully number off correctly in the front rank ... or we would just carry on numbering off in the second rank when the front rank was done.
>
> At night, too, the orderly officer used to make the rounds of the bungalows to make sure that we were all present and tucked into bed. We started off by putting dummies under the blankets of those who were absent but the inspection party soon cottoned on to that, so we would move stand-in volunteers to fill up the first two bungalows just before inspection and then

leap-frog along the line as they finished inspecting each bungalow in turn. Then someone in authority had a brainwave; on parade one morning we were given special disciplinary powers to keep us under control. The names were then read out and they turned out to be the biggest lot of troublemakers amongst us. They took their newly gained authority very seriously and knowing the ropes as they did, succeeded in improving discipline considerably.

Dad had come to Tempe as a sergeant in the Air Force, and was able to keep his rank. This was beneficial at the time, because sergeants received reasonable pay and he was able to allocate some of it to his mother and sister, Sheila, who were living in Pretoria with relatives because Pop had already been posted up to North Africa with the first group of South African Air Force personnel to be mobilized.

From Tempe they were sent down to Durban, to the SAAF Electrical and Wireless School on the Marine Parade. Here they learned Morse code and how to handle radio sets. According to Dad, 'The air gunner training was pretty general. You were expected to become a wireless operator, to handle guns, later to do navigation and be a bomber as well, so we got a thorough theoretical training in all these things.'

In their time off they all had a whale of a time in Durban. Bob Rogers remembers:

> We were paid very little in those days. I think 2/6 per day (25 cents), but in Durban three or four of us would pool our resources and hire a car on a Saturday night and take our girlfriends dancing.

They left Durban, all conversant in Morse code, and were then sent to East London. They had finished their ground training, and were happy to start on their flying training as air gunners. They were elated. On arrival in East London their hearts sank as there were over 700 young men waiting to go out for flying training as air gunners, and every three weeks another eighty to ninety new trainees would arrive. The trainees were going up for training at the rate of five every six weeks. The SAAF did not have enough aeroplanes, and the resulting build-up of impatient young men, all confined in one place with nothing to do, put paid to discipline. Dad chuckles, 'They just couldn't do anything with us. Discipline went out of the window. We were prepared to work; we were prepared to do anything, as long as we were part of the war. But this seemed to us the biggest bloody waste of time under the sun ... and the things we did! We used to quite blatantly grab our swimming trunks, wrap them in a towel, and go round in full sight of the guards, jump the fence and bugger off to the beach for the rest of the day. We thought, to hell with roll calls and parades. And there were so many of us breaking rules that those in authority had to just turn a blind eye to it all.'

One day, the trainees were being marched back from the shooting range. 'Don Tilley, who had a peculiar sense of humour, was marching in the front middle row. He whispered to the chaps on either side of him, "Pass back quietly and when I shout 'snake', you all fall out and rush into the bush." So they passed the word down, while the sergeant next to them was shouting left ... right ... left ... right. Suddenly, Tilley jumps up and shouts, "SNAKE!" at which the whole squad disappeared into the bush. The poor sergeant was doing his nut trying to get some order back into this lot. Typical Tilley humour, when the sergeant asked who had started it, Tilley says, "Well Sergeant, it was this chap," (pointing to someone else) "and I don't think there was a snake there at all!" So the poor bloke was arrested and put in the guardroom until we all went to the sergeant and said, "Sir, you can arrest all of us because we were all involved." Tilley thought it was a helluva joke.'

One evening, the orderly room clerk in the Course 7 bungalow said, 'Look chaps, you had better be at roll call tomorrow morning; something's cooking. I heard something in the adjutants' office today.' Wondering what this something was they passed the word around quickly. The next morning they were all standing in anticipation on parade. The adjutant addressed them: 'We have had a request from the Royal Air Force for volunteers to train as pilots. If you volunteer you will have to go in front of the RAF selection board. If you are accepted, you will be seconded to the RAF, and go to Southern Rhodesia for pilot training. When you are dismissed from parade those wanting to volunteer are to come to my office to give me your names.' Dad characteristically describes what followed: 'God! What a bloody stampede to that office!'

Chapter 8

Taking Flight

According to Roy Conyers Nesbit in his book *Torpedo Airmen*, the sudden call from the RAF for trainee pilots was due to 'the loss of 250 trainees from Britain, torpedoed in convoy in the Atlantic.' Dad was one of the 250 hopeful volunteers who were all sent up to CAFTD to go before the selection committee. In December 1940, he went in front of the RAF panel and was accepted on condition that he had the consent of his parents, because he was not yet twenty-one. As soon as they were dismissed, he got permission to go to Pretoria to ask his mother for her consent. He tore into town, and after a hurried reunion with her and his sister, he excitedly told her what had transpired and said, 'Mom, please, you know I have always wanted to fly. This is my chance. Please, please, please!' She signed her consent and he returned to camp. The volunteers who were accepted were told that the first group would go up to Southern Rhodesia in the next few days, and the second group would leave the following week. The first group were selected, and duly left for the RAF training base at Khumalo, in Bulawayo. In spite of his pleading, Dad was not among them.

The following week arrived and nothing happened. Another week went by and still nothing happened. This resulted in a build-up of tension, and the boys started getting very stroppy. They were on the verge of mutiny because they had been messed around so much. Eventually a few spokesmen were selected from the group – Dad was one of them – and they requested an interview with the CO. It was granted and they went to him and said: 'Sir, with all due respect, and without any intention to be insubordinate, we have just had enough! We have been in the Air Force for a year now, we have been pushed from pillar to post, and we have been promised things that don't happen. We have been here, there and everywhere, we have finished our basic training, and now we are being pushed around like a bunch of rookies. We don't want to cause any difficulties, but please, can't you see your way clear to give us some leave until we are called up?' He saw reason and agreed to give them leave, on condition that they were to give the addresses of where they would be staying in Pretoria, and be on twelve hours' standby. It was now the middle of December and the beginning of the Christmas holidays.

They were all by now veterans at ways and means of bucking the system, or so they thought. Dad, Wally Morgan, Dick Hague (who later became an air traffic controller at the then Jan Smuts Airport) and another chap by the name of Miller got together and decided they would go down to Durban, a 500-kilometre trip. The reason was that Wally and my father had girlfriends in Durban, and the other two just fancied the trip. 'Miller had his father's car, and although we were on twelve hours' standby, we reckoned we could just make it back in time if we were called up. So we made elaborate plans with our mates to use addresses and phone numbers in Johannesburg, and they agreed that if they were called up they would immediately phone us in Durban. And off we set.'

They had only been in Durban for three days when they got a message from one of the blokes. 'Everybody has to report to camp at nine tomorrow morning.' Dad, laughing at this predicament, says: 'God, what a flurry and scurry. We rushed around getting all our kit and all the blokes together and into the car, and off we set on the road back to Johannesburg. Well, at about midnight we got to Newcastle and the bloody car broke down. We were stuck, with a broken half shaft on the car, between Durban and Jo'burg, and boy, we were all in a hell of a panic!'

They quickly made some enquiries and found out that the night train came through at two in the morning, and managed to find some bloke who would fix the car and so left it with him. When they threw all their money together there was just enough for their fares, so they then hurtled through town on foot to catch the train. As they leapt onto it, the Afrikaans conductor threw them off because it was fully booked. They then ran down to the dining carriage and jumped on there, only to be confronted again by the furious conductor who shouted, '*Julle skelms! Julle mag nie op dié trein ry nie, julle moet afklim.*' ('You rascals, you cannot stay on the train, you must get off.')

Dad tried in his best Afrikaans to explain their urgent situation. '*Asseblief, meneer, ons is opgeroep en moet vanoggend in Pretoria wees.*' (Please, Sir, we have been called up and must get back to Pretoria by morning.)

The conductor thought for a moment. '*Nou goed dan maar julle kan nie hier bly nie. Volg my.*' (Well, all right, but you can't stay in here – follow me.) He led them to a grubby third class compartment where they spent the rest of the night, and the next morning they alighted in Germiston and caught the suburban train from there to Pretoria Station, arriving at half-past eight. Miller phoned his father, who, although very unimpressed with their foolishness, collected them and drove them to camp, getting them there dead on the dot of nine.

They bolted into the CO's office and stood to attention, saying in unison, 'Reporting for duty, Sir!'

He gazed at them blankly. 'What are you doing here? You're supposed to be on leave.'

'We got word to report back at nine o'clock this morning, Sir,' they replied.

He assured them, 'Well, there was no such order given, so you can push off again!'

'Yes, Sir, thank you, Sir,' they all said, trying to maintain their composure. When they got out of the office, they discovered that the only person required at the camp that morning, for some administrative matter, was the mate who had phoned them. Dad laughs, 'Oh God, after all that palaver and carry on! Anyway, we all clubbed together and paid for Mr Miller's car to be repaired and returned to Johannesburg.' They spent the rest of their leave hanging around camp, kicking their heels.

The year 1941 arrived and in early January the second group of trainee pilots were fitted out with their flying kits. They were finally dispatched to ITW (Initial Training Wing) to join the first group at Khumalo Air Training Base in Bulawayo, Southern Rhodesia. Here, in addition to being with all his close trainee friends, Don Tilley, Gill Catton, Bobby Rogers, Wally Morgan, Gordon Brodziac, Jack Lever, Hugh Sheldon, Cliff Evans and others, Dad was reunited with his old school friend from Pietersburg, Raymond de Marillac.

On arrival at Khumalo they found that even though they had all already done a year of drill, physical training and lectures, they were expected to do it all over again. This didn't go down well with them, so the new SAAF trainee pilots worked on a ploy for their next drill session. When they assembled on the parade ground with their drill instructor, Sergeant Bennett, they raised their protest with a proposition. Their spokesman stepped forward and said, 'Look, Sergeant, we can do all the drill you want, and we

A group of excited trainees at Lyttleton, Pretoria, before leaving for Bulawayo. Second row, third from right, Ted, pleased as punch, and Don Tilley, far right.
Author's collection

can do it better than any of you can teach us. We want to prove this to you. May we have permission to show you right now? And please, after this, we don't want any more squad drill.' Sergeant Bennett, obviously intrigued, agreed to their request.

Dad, being a well-practised drill instructor, was selected to give the commands for an advanced drill they had learned in East London. On his command, they proceeded. Like a ghost squad, they began with absolute precision to move as one body in perfect synchronized motions and enact what my Dad called the 'silent drill'. He described it thus: 'I would give a softly-spoken command, no shouting, and the squad would move accordingly. Stepping softly, and taking hold of their weapons without slapping their hands on them, and placing the butts silently on the ground. There was no stamping of feet, for instance on the command of "halt", instead of ONE, TWO, CRASH as they brought their feet down, it was just a soft step, step … and dead silence.' By comparison with the normal jarring, crude foot-stomping drill, this had a compelling beauty to it.

TAKING FLIGHT

While watching, Sergeant Bennett had been joined by the sergeant major, and they were indeed very impressed because, as Dad says proudly, 'The boys were good, make no mistake; when they put their minds to it they were fantastic!' When they had finished their routine, Dad said, 'Look Sir, we have finished with this lot, we know our stuff, and don't want to keep doing it!'

He replied, 'Fair enough chaps, that's not unreasonable,' and after this the boys had an easier time of it, and the physical drill requirements were relaxed.

In a letter to me, Bobby Rogers comments on their uniforms and their instructor:

> Sergeant Bennett wasn't very impressed with our short shorts, which he considered to be rude and almost obscene (the RAF wore their shorts two inches above the knee). This led to further arguments as we were instructed to replace our existing shorts immediately, which we refused to do.

Apparently, other RAF instructors on their course also had a difficult time with the rebellious behaviour of, as Dad sums them up, 'this bloody raw, rough crowd of South Africans'.

The instructor who took them for English, as Dad describes him, 'was the honourable son of so and so, a typical, rather affected and slightly effeminate English aristocrat.' Bob Rogers writes:

> Our rather effeminate English instructor was Lord Malcolm Douglas Hamilton, one of four quite famous brothers, all of whom served in the RAF. We did lead the poor guy a merry dance. If I remember correctly, he was also supposed to teach us navigation, but we usually managed to get him to talk on something else. We early on asked him if he was related to the royal family and this led to endless explanations of how they (his family) fitted in and endless stories of his family and friends and their castles etc. ...
>
> One day in class he was almost in tears, saying that we were dreadful people and asking why we couldn't keep quiet for a change. We told him we could do so if we chose. He challenged us and said, 'All right, let's see how long you can keep quiet. One, two, three, go!' At which, we all got up and started walking out the room.

'Stop. Where are you going?' he cried, running up and down trying to turn us around. 'You told us to go, Sir,' we said. 'No I did not!' he shouts. 'Yes, Sir, you said, "One, two, three, GO!"' we replied. Poor chap! It sounds so stupid in retrospect but we thought it was hilarious at the time. I often felt quite sorry for him but in the end I think he had a soft spot for us despite the trouble we caused him.

One day, the chaps found out it was their instructor's birthday, and during their lesson they stood up as one man and, at the top of their voices, burst forth loudly with, 'Happy birthday to you, happy birthday to you …' He was only a sergeant and so was running up and down the classroom saying, 'Shhhhh … quiet, the CO is over the road, quiet … shhhh.' This noisy and boisterous bunch of South Africans, in the face of pukka, gentlemanly British restraint, was a trifle too much to handle. The mischief making was, of course, all in the spirit of harmless fun and stemmed from the boredom of base camp life. The authorities seemed to understand this, which resulted in the building of a good rapport between them and the pupils.

Then, eventually, THE BIG DAY arrived when they were to move over, from the Khumalo ITW camp to EFTS (Elementary Flying Training School) at Induna, where they would, at last, start learning to fly.

Induna Aerodrome is situated on an open area of Bushveld near Bulawayo.

It is named after the historic flat-topped hill called Ntabazinduna, which rises from the expanse of surrounding veld, as a solitary landmark. On the day they moved to Induna, with boyish excitement they proudly donned their flying kit, which consisted of a brown leather lumber jacket with a removable brown fur collar, a close-fitting black leather helmet that buckled under the chin, and flying goggles.

As soon as they arrived, they went to look at the aeroplanes parked on the runway. Dad passionately recalls this day: 'We were all kitted out for the first time in our flying gear, to look at those beautiful aeroplanes, the lovely old Tiger Moths.' Laughing, he says, 'Of course, we were parading around making out we were big deals and heroes, all taking pictures of each other, standing proudly next to the Tiger Moths, never having even been up in one. With me at the time were Bobby Rogers, Wally Morgan and Cliff Evans.[1] We all stuck together and had become very close mates.'

At last, in January 1941, they started flying. Pilot Officer Dereck Chidell was Dad's flying instructor. 'He was a super chap, he roared me around the sky in this Tiger Moth, doing loops that turned the world and my stomach

TAKING FLIGHT

Tiger Moths lined up on the runway at Induna, 1941. Author's collection

Ted posing happily on a Tiger Moth, Induna, 1941. Author's collection

Course 3B 27 EFTS, Induna, Bulawayo, March/April 1941. Top row, from left: Doyle, SAAF, Hartley, SAAF, Hore, SRAF, Young, SRAF, McGlashan, SRAF, Flote, SRAF, Biddulph, SRAF, Willis, SRAF, Greyvenstein, SRAF, Launder, SRAF, Du Preez, SRAF, Duigan, SRAF, Humhreys, SRAF. Second row, standing: Evans, SAAF, Berry, SAAF, Wolf, SAAF, Lazarus, SAAF, Knight, SAAF, Van Dyck, SAAF, Cohen, SAAF, Van Rooyen, SAAF, Strever, SAAF, Glanville, SAAF, Flemington, SAAF, Driver,

upside down, and stalls where all went silent, and spins that made me dizzy; it was truly exhilarating.' At ETFS they were trained in and practised aerial exercises, starting with circuits and bumps (take-offs and landings), then on to side-slipping to land, stalls and spins, steep turns, and cross-country flights, which included navigation training.

TAKING FLIGHT

SAAF. Third row, seated from left: Dorner, SAAF, Mann, SAAF, Donald, RAF, Douglas, SRAF, Dawson, SAAF, Hughes, SAAF, Wilson, SAAF, Davies, RAF, Lyon, SAAF, Morgan, SAAF, Rogers, SAAF, Warren, SAAF, Caithness, SAAF, Cummings, SAAF, Wilbur, SAAF. Front row, from left: Judson, SRAF, Richardson, SRAF, Ledger, SRAF, Barbour, SAAF, Hastings, SAAF, Darryl, RAF, Brodziac, SAAF, Unknown, Boswell, SAAF, Richards, SRAF. Author's collection

In *Heck! What a Life*, Carey Heydenrych writes that there was also instruction in 'aerobatics, formation flying, low flying and night flying'. When this training was completed, the next big step was – going solo.

In February, Dad celebrated his twenty-first birthday at Khumalo, with a flattened cake and, I'm sure, a few 'toots' with his mates. His mother had sent him, all the way from South Africa, a beautifully iced cake, which had a little aeroplane standing on the top of it. But having travelled through the postal service, by the time it reached its destination it was fairly well squashed; the little aeroplane was thoroughly grounded, embedded in the icing like a bad omen. (In spite of its unappetizing appearance and the negative analogy, it was thoroughly enjoyed.)

As camp life wasn't all about 'circuits and bumps', there was an old His Master's Voice wind-up gramophone in the barracks. It had a brass trumpet-speaker displaying the unforgettable logo of a fox terrier with his ear turned to the music. As it needed new needles, Dad and his old school friend Raymond De Marillac[2] went into Bulawayo one day to obtain a packet of them. They entered the music shop of Laurence & Cope and approached the counter. A tall, elegant girl with lustrous dark hair emerged from the office to assist them. After buying the packet of needles, they stepped out of the shop and Dad turned to Raymond. 'Boy, oh, boy, what a honey!' He was smitten, and characteristically states: 'And that was that!'

Chapter 9

Cloud Nine

Pilot training for Dad was henceforth punctuated with distracting thoughts of the striking but mysterious girl who had sold him a packet of gramophone needles. Luckily, a few weeks after the music shop encounter, the trainees got word of a forthcoming dance event in Bulawayo. A group of them managed to find partners and, Dad says, 'Off we went to the dance, and who should be there, with someone else, but the lovely honey from the music shop. During the evening I asked her to dance, and I'll never forget the dance I had with her that night. Oh, it was wonderful. As she wasn't my date I, of course, couldn't take her home that night. I was furious because I really fancied her, but at least I now knew her name was Bee!'

Bee, Bulawayo, c.1944. Author's collection

A few days later, one of the trainees asked Dad to take a message to his girlfriend in town as he was on guard duty and unable to leave camp. Dad immediately saw this as an opportunity to somehow visit Bee at the music shop again. He chuckles, 'I thought, this is crafty, I'll pretend I don't know where to find this girl. I'll go to the music shop and ask Bee if she can help me.' He laughs at himself and says, 'Oh, it sounds so corny now, doesn't it?' Once in town he went into the shop and asked Bee if she knew where he could deliver the message. She gave him the directions he needed, and as he was leaving the shop, he turned around and asked, 'Oh … um … by the way, would you like to come swimming with me on Saturday afternoon?' In spite of this less than enchanting and unromantic invitation, she replied (in her very good English diction): 'Yes, I would like that very much.' He

Pop and Granny Clarke, Bulawayo, 1940. Author's collection

left with a first date in his pocket, and on Saturday afternoon Dad and Mom had their first date at the Borrow Street municipal swimming baths.

My mother's family lived together in a small house on Wilson Street in Bulawayo. Granny Clarke, my maternal grandmother, was born in Swansea, Wales; was a staunch Catholic; had a bit of the Welsh dragon in her; and ruled the roost. Pop Clarke, my mother's father, was born in Sussex, England. He had come out to South Africa during the Boer War as a 'horsehand' on the ships ferrying horses from Britain to South Africa. He was a gentle, sentimental old soul, whimsical and given to dreaming. My mother would tell us, always with a tear in her eye, that, 'In the evenings he would sit out on the veranda, put me, his "little Bertha", on his knee, and sing

CLOUD NINE

Under the Old Apple Tree and *After the Ball is Over*.' It was one of her most enduring childhood memories.

Dad said, 'I got on very well with Bee's family. Old Pop Clarke, Bee's father, was a real softy and occasionally I would entice him to have a couple of beers with me at the old Victoria Hotel, which was around the corner from their house in Wilson Street. At the time the hotel was a wood and iron structure in the true pioneer style, with a wooden veranda. We used to sit there on a bench on the veranda and have our beers. Beer was nine pence a pint then. But Ma Clarke, as I got to call her, was a real tiger and I had to get around her by joking with her or teasing her. Pop Clarke was a very skilled artisan in mosaic work, tiling and building, because you know, in 1914, he laid the marble tiled floor of the foyer to the Johannesburg City Hall and did the mosaic work in the Wolmarans Synagogue, and they remain there today as a testament to his workmanship. He didn't have a job at that time and the family was very poor. They relied totally on their daughters' income.'

He did, however, have the little gold mine called Serten, near Thaba Nyaka, about 20 kilometres southwest of Bulawayo.

Bee, second from right, at the Serten Mine, near Bulawayo, late 1930s. Author's collection

ON LAUGHTER-SILVERED WINGS

My father takes up the story …

Once when I had a few days' leave I went out there with him to check up on things. All that was there was a little wood and iron room, which we camped in for a few days. Pop Clarke and Ernie, Bee's younger brother, had started developing it. They had started by digging a vertical shaft straight down into the ground, like a deep well, and from the bottom of it they had dug a tunnel

Ernie standing with Pop Clarke, and Bertha ('Bee') seated on the Chevrolet, c.1935.
Author's collection

CLOUD NINE

running horizontally. Things were very rudimentary in those days. The means of getting up and down the shaft was a zinc bucket tied to the end of a rope. Swinging from side to side, with one foot in the bucket, and one foot kicking away from the sides, you were jerkily lowered by hand down the shaft.

However, when I met Pop Clarke the mine wasn't working, as four years earlier tragedy had struck the family. While Ernie was working there, his African worker fell down the shaft that they were both digging. To rescue him Ernie reached down, gripped the worker's arm, and with great effort physically pulled him out. The terrible strain of this had torn the lymph glands under Ernie's arm, the injury developed into cancer. Within weeks the family lost their only son and brother. Ernie died at the age of twenty-one. Bee, aged sixteen, was cradling him in her arms when he died, and the whole family was heartbroken. Ever since then Pop Clarke's heart just wasn't in the mine any longer.

In his yard Pop Clarke had a dilapidated 1929-30 open tourer Chevrolet. It had been fairly well hammered going through the bush to the mine for years, and was pretty far-gone. The windscreen was held up by a piece of wire tied to the bar on which the front headlights were mounted. The tyres were just another story … I mean, if there were traffic officers in those days no ways would it have ever been allowed on the roads. Pop wasn't interested in driving it, and Vi, who could drive, refused to drive it. Well, you know me and cars! I offered to take it out to the camp and get it fixed up, which Pop happily agreed on, and we had great fun in the old jalopy. Oh God! One sweltering day Bobby Rogers and I were driving into town from camp all smartened up, and we had a puncture. I remember the two of us sweating and straining in the heat to change this bloody tyre, us both ending up grimy and dishevelled.

I, of course, used to go into Bulawayo to see Bee as often as I could and we fell very much in love. At this time Bee was training in her spare time as a classical singer under the superb tuition of Elsie Fraser Munn. To bring in some extra income for the family, she sang in the Grand Cabaret at the Grand Hotel every Saturday night. The car enabled us trainees to go in on the occasional Saturday night to enjoy an evening of music, dining and dancing. The manager, Bill Hardy, was fantastic to us. We sat at his table and he would never let us pay for a thing – thank God, because we only earned seven and six a day.

One night after the cabaret we decided to go out to the Round House Hotel about 10 miles out on the Waterford road. We were very young and this felt quite daring as it didn't have a good reputation (I don't know why because actually it was quite respectable; perhaps it was because it was so far out of town). Well, we stayed there until about 1.30 am and, lo and behold, when

ON LAUGHTER-SILVERED WINGS

Bee in stage costume, c.1940. Author's collection

we came out we had a puncture. We arrived home at three o'clock to a furious Ma Clarke, who was waiting for us. She shook her finger at me and said, 'Don't give me your excuses, don't you dare bring my daughter home at this time of night!' Oh Lord, did I get it in the neck from her!

On another occasion, when I was doing night flying training on Oxfords, my instructor unofficially allowed me to beat-up [repeatedly fly over at low altitude] their house ... there were no high rise buildings then and we had quite an exciting time buzzing around the house. Those were happy days.

Chapter 10

Taking the Controls

In the meantime, training on Tiger Moths continued at Induna, until the time came to go or not to go solo.

There was great competition to see who could go solo first and what an exhilarating feeling it was to be in the air at the controls of your own Tiger Moth.

> After a few hours we thought we could really fly and issued challenges to each other. 'Meet you over the dam at 2,000 feet.' We'd have a long white scarf around our necks and proceed to dogfight and show off. How we didn't meet with nasty accidents I don't know. Ignorance is surely bliss.
> Lieutenant-General Bobby Rogers, in a letter to the author

The excitement of going solo was always tempered by some tension as trainee pilots were sorted into two categories: passed or failed. The ones who didn't make it were 'washed out' of the Air Force by being what is called 'bowler-hatted'. This was quite callously and unceremoniously done by the instructor, who, on the flight officer's notice board, merely drew a bowler hat (which symbolized the inevitable return to civilian life) next to the name of the unfortunate person. This happened to about 50 per cent of pupil pilots. When it happened it was a sad time of goodbyes, because firm bonds of friendship had been created.

Dad passed and was finally able to go solo. 'Going solo,' he says, 'was the start of a new life, and a new way of experiencing the world. It was thrilling!'

> Here we go. I release the brakes. There is no instant rush of speed, no head forced against the headrest. I feel only a gentle

Ted going solo in a Tiger Moth, Induna, 1941. Author's collection

> push at my back ... (and as the wheels leave the ground) ... there is nothing in the world but me and an airplane alive and together and the cool wind lifts us to its heart and we are one with the wind.
>
> Richard Bach, from *Stranger to the Ground*

Dad passed through his training on the Tiger Moths very successfully with an above average rating. Once the trainees had qualified as pilots, they were put onto either twin-engined or single-engined aircraft. This unfortunately meant they were sent to different airbases, and so again there were goodbyes as friends' paths diverged. Dad explains: 'This is when Bobby Rogers and I parted for the time being; he was sent up to Cranbourne SFTS at Salisbury, to fly the single-engined Harvard, and I returned to Khumalo to fly Oxfords.'

Until this time Dad had been very proud and happy with his flying training record, but just at the end of advanced training, 'Oh Lord, disaster struck! I was on an instrument flying exercise with Willy Wilson.[1] He was

TAKING THE CONTROLS

doing the instrument flying and I was responsible for the aeroplane. I was seated in the right-hand seat of the aeroplane – which I was unaccustomed to – and when landing the aeroplane, in a moment of confusion on touchdown, I pulled the bloody undercarriage up; the alarm hooter sounded and I realized what I had done and pushed the lever back, but it was too late and POWWW ... down we went. One wing and the propeller smashed into the ground. I put my hands to my head and thought, oh God, no! Just when I am about to pass out! I was devastated.'

After recovering from this dreadful accident Dad had visions of being bowler-hatted from the Air Force and the shock of the entire ordeal left him badly shaken, 'not to mention the shame and chagrin of having done such a stupid thing. Of course, I got a lot of ragging about it from the boys, which didn't help. I'll never forget that weekend. I went to the Grand Cabaret with Bee and I was a real Dismal-Daniel.'

A few days later, he was called up before the CO of the Advanced Training Squadron, Squadron Leader Hallmark, who severely reprimanded him for 'this act of negligence', and had his logbook endorsed with the word 'careless', but in spite of this huge error of judgement he was permitted to complete the flying course. Soon after this, and upon completion of their training, all the cadet officers were called up by their commanding officer, Group Captain Dalzell, and in a ceremony under the historic Indaba Tree at Khumalo, all were congratulated and highly praised and commissioned into the Royal Air Force as pilot officers.

Dad had achieved first place in the wings examination of the course. The new pilot officers were instructed to put their blue 'kings and wings' on their uniforms. Dad says, 'I was extremely proud of becoming a pilot officer in

Ted proudly wearing his RAF uniform, Induna, 1941. Author's collection

ON LAUGHTER-SILVERED WINGS

July 1941 in the Royal Air Force wearing the Royal Air Force blue. As the Battle of Britain had peaked at the end of September 1940, and the RAF took credit in downing 240 enemy aircraft, it was considered an honour to be a member of the RAF during this historic time.' With great excitement he immediately went, in his new blues, to Bulawayo, to give Bee the good news and he asked her to sew his blue 'kings and wings' onto his uniform. (Before she sewed them onto the front of his jacket, she secretly embroidered the words 'God Guard You' on the back. Dad was unaware of the little secreted prayer that he wore over his heart throughout the war.)

The RAF wings, secretly embroidered on the back by Bee and sewn onto Ted's uniform in 1941. Author's collection

After being commissioned, all the graduates were due for leave. However, the RAF called for volunteers to undertake the job of ferrying nine or ten Oxford aeroplanes from Stamford Hill (now the Kingsmead Cricket Stadium) in Durban to Khumalo in Bulawayo. Wally Morgan, Willy Wilson

TAKING THE CONTROLS

and my dad were among those who volunteered to undertake the trip. They were dispatched by rail to Durban, all smartly dressed in their uniforms and in high spirits.

They arrived at Stamford Hill excited at the prospect of making their first long-distance solo flight. Warrant Officer Adkins AFM (Advanced Flying Medal) was to lead them in loose formation to Swartkop Aerodrome at Pretoria and then on to Khumalo. On take-off, recalls Dad, 'The Oxford took off like a rocket from Stamford Hill; it fairly bounded into the air, much to my surprise and consternation. I was amazed at how quickly we were able to reach the required altitude. All my previous flying experience at Bulawayo had been done at the altitude of about 4,000 feet, whereas here we were at sea level and the plane had much more power.'

As the loose formation of aeroplanes flew past Ladysmith the weather deteriorated and, without warning, Adkins, the leader, who was flying about 50 feet above the ground, did a steep turn and returned to Durban. Dad laughs, 'The gaggle of low-flying aeroplanes all trying to avoid him and then doing steep turns in pursuit of him – it was quite hair-raising and funny at the same time.' Back at Stamford Hill they waited for more favourable weather and the next morning they took off again. They successfully made the flight to Swartkop and spent the night in the officers' mess.

Of the briefing the next morning, Dad says, 'I was astounded at the casual manner of briefing. There were no maps or provisions given to us, and no radio communication between the pilots. So in the event that a pilot was unable to keep up with the formation, or lost it in cloud, he was totally alone, and would have to navigate himself back to Khumalo.' They took off. As they neared Warmbaths, cloud started to build up, so the formation climbed above it. 'When clouds started forming we climbed above them, then the leader indicated to us by pointing his hand down that he was going to go down through the clouds and that the company of aircraft should follow. Needless to add, this was extremely perilous to the rest of the pilots who found themselves in dense cloud with other aircraft in close proximity. I didn't like this at all, because the moment we were in the cloud I lost sight of him, and no ways was I stooging around in the clouds with other aeroplanes around. This was definitely for the birds!

'So I decided to pull up above them and continued on the compass bearing we had been flying on, not knowing how long it would take me to Khumalo. Somewhere over the Matopos Hills I found a gap in the clouds and did a spiralling dive through it. This left me totally confused. I didn't know where the hell I was! But I continued to fly on the compass course, with the nagging worry about the limited fuel I had to reach the destination. Eventually I came across a railway line and by guesswork followed it eastward and finally landed at Khumalo. All the planes arrived safely after a somewhat confusing flight, coming in from all points of the compass.'

From Khumalo, having passed through SFTS (Secondary Flying Training School) most of the pilot officers were selected to fly with RAF Coastal Command and were posted to the RAF School of General Reconnaissance at George, in South Africa. Dad, however, was posted to the Central Flying School in Bloemfontein, as a flying instructor. 'It absolutely horrified me!'

Chapter 11

The Big, Wide World

> *When you take your journey to Ithaca*
> *Then pray that the road is long*
> *Full of adventure, full of knowledge ...*
> C.P. Cavafy, from the poem *Ithaca*

Although he was aware of the huge responsibility and vital role of a flying instructor, for my father it was like being condemned to the role of an onlooker. This was definitely not in keeping with his character.

'Luckily,' he said, 'I knew a chap called John Ledger, who didn't want to go on to navigational training. He wanted, rather, to be an instructor.' Together, Dad and John appealed to the CO to switch their postings and, to their mutual relief, this was approved and implemented. With the promise of active participation, he happily joined his buddies in No. 13 NR War Course, who were set to depart for George.

Before leaving for George in August 1941, the pilot officers were offered the chance to transfer back to the SAAF as second lieutenants or remain in the RAF. Dad explains: 'The SAAF pay was better than the RAF's. We would earn twenty-five bob a day plus five bob a day flying pay, whereas the RAF pilot officers earned fourteen shillings and sixpence a day. So we all went back to the SAAF, which meant we also received our SAAF wings, and were then seconded to the RAF.'

In George, most of the trainees who had been separated in Southern Rhodesia for different flying courses were reunited. Dad was very happy to be once again among his trainee friends Don Tilley, Wally Morgan, Willy Wilson, Gill Catton, Carey Heydenrych, Hugh Sheldon and Jack Lever. In the navigation course at George, they flew in pairs in Avro Ansons with South African staff pilots (referred to by them as 'stooge' pilots). Here they became well trained in astronavigation and general reconnaissance and were put through the paces of being first and second navigators, the first navigator

Group No. 13 of Navigation and Reconnaissance Course, taken at Induna before leaving for George, 1941. Ted is seated second from left. Author's collection

A training exercise in an Avro Anson with a 'stooge' pilot, NR Course, George, 1941. Ted is on the right. Author's collection

THE BIG, WIDE WORLD

Avro Anson in the air, George, 1941. Author's collection

executing the plotting at the chart table, the second navigator working on a tiny board on his lap.

In *Heck What a Life!*, Carey Heydenrych recalls:

> We would spend the next three months undergoing intensive training in some fascinating specialized navigational practices. Basic instruction took place on the ground in the classroom but our practical training took place over the sea in twin-engine Anson aircraft. We, the pupils on the course, did not fly as pilots but only as navigators. Our navigational exercises were conducted far out to sea, about 100 miles or more from the coast, and well beyond the sight of land. On a clear day, George Peak, situated behind the town of George and the most prominent peak in the Outeniqua Mountain range, came into view from about 80 miles out to sea. The sight of this grand peak from so far out at sea was a welcome one. But to the newcomer to sea navigation it never seemed to appear where he expected to see it; it seemingly materialized on the horizon to tempt uncertain navigators to doubt the accuracy of their dead reckoning calculations and to alter course to aim for the peak. Those who succumbed to the beckoning tip of George peak in the far distance and turned towards her, would make a bad landfall when crossing the coast, and end the exercise miles off course.

One of the other elements of the training exercises was to investigate any ships spotted out at sea. This meant 'beating the ships up' in order to identify them. Dad describes one of his more unpleasant experiences: 'I'll never

Ted working at navigation on the NR Course in an Avro Anson, George, 1941.
Author's collection

Ted at George airbase with George Peak and the Outeniqua Mountains in the background, 1941. Author's collection

forget. We had a party one night and the next day I was feeling far from strong. We went out on an exercise and saw ship after ship. The pilot was having the time of his life roaring down over these ships. Oh God, it was awful! I ended up at the chart table as first navigator, passing up the course, then running over to the window, heaving, running back, passing up the next change of course, then running to the window again.'

In spite of some revelling, when this training was completed the CO again gave Dad a good report for his navigational work. Course Commander Squadron Leader Harrison then called them into his office and asked, 'Your group is to fly in Coastal Command in England. Do you want to be on flying boats, do anti-sub patrols in Hudsons or be in Bomber Command with torpedo bombers?' Almost everyone's reply was 'torpedo bombers'. 'Because,' as Dad says, 'they sounded much more adventurous and everyone was looking for some sort of action. However, Squadron Leader Harrison, who himself was a flying boat bloke and believed them to be the cream of the Air Force, thought their choice absurd, and said, "You're crazy, there is no future in that!"'

A few days' embarkation leave was then issued for the group to see their families before leaving South Africa for Great Britain. Dad spent the time in Johannesburg with Bee, who had travelled down from Bulawayo, and also with his family. After a few days together they said their farewells before he boarded the train for Cape Town in October 1941.

In Cape Town, the pilot officers boarded the P&O passenger liner, HMS *Strathnaver*. They were the first contingent of South African airmen to leave for service outside. The *Strathnaver* had been converted into a troopship and was returning to England after having delivered troops to the Middle East. 'We were very lucky,' Dad says, 'it was fairly empty and didn't have the thousands on board that it usually carried, so we found good accommodation on the first deck. Onboard we experienced blackout conditions for the first time.'

> At night, we would sit out in the open on the top deck singing. … No smoking and no lights of any kind were allowed on deck. In the tropical heat, the smell of the diesel oil being burned by the ship's engines was oppressive but we nevertheless enjoyed the fellowship. … Near the equator the phosphorescence churned up by the ship's propeller wake was quite spectacular.
>
> Carey Heydenrych, from *Heck What a Life!*

Ted onboard the Strathnaver, *Nov-Dec 1941.* Author's collection

THE BIG, WIDE WORLD

As the young pilots were the only military personnel onboard they were briefed on how to man the guns in the event of an attack.

> We [had] set sail from Cape Town on our own and without escort. We were armed, albeit poorly. Our armament amounted to three machine gun relics of World War 1, mounted in a mock funnel forward in the ship, and the other two in lookout nests mounted high above the deck. In addition to these machine guns, a medium-sized naval gun was mounted astern to fend off attacks from the rear. An anti-aircraft balloon, tethered above the stern of the ship was there to discourage low-level aerial attacks. We pilots were deemed to be the best trained of the personnel aboard and were selected to man the guns. We were split into two teams, working in shifts, to cover these duties.
>
> Carey Heydenrych, from *Heck What a Life!*

The ship headed across the Atlantic for Port of Spain, Trinidad, all the while zigzagging wildly to avoid enemy submarine attacks. She was due to proceed from Trinidad to New York, before crossing back across the Atlantic to Greenock harbour on the Clyde River, near Glasgow, in Scotland. However, on the twelve-day trip to Trinidad, orders were issued to cancel the stopover in New York and to proceed directly from Trinidad to Greenock. Apart from this news, the trip to Trinidad was uneventful. In between manning the guns, which on the way to Trinidad was uneventful. In between manning the guns, which on the way to Trinidad was 'a piece of cake', a fair amount of fun was had on board with deck games, cards, singing and pink gins. After being confined on the ship for twelve days the arrival at Port of Spain was welcomed with much excitement. At lunchtime the ship slowed down to an almost motionless glide and slid gently towards the narrow passage of the southernmost entry point to Trinidad. From their seats in the dining room the passengers watched in awe as:

> In utter silence and almost motionlessly, she glided through the narrow gap between the steep sides of the entrance passage – a breathtaking and unforgettable experience.
>
> Carey Heydenrych, from *Heck What a Life!*

ON LAUGHTER-SILVERED WINGS

The *Strathnaver* was too large to berth in the Port of Spain harbour, so she anchored in the roadstead, and passengers were ferried on a 'lighter' to and from shore over the next two days.

After anchoring for the night, the young pilots were allowed to go ashore. Dad says, 'We walked around and had a look at the port, did some shopping and posted off postcards to our people back home. During our meanderings we came across some rather lovely girls, and one of the brasher types among us said, "Could we perhaps go dancing tonight, we've come from South Africa … etc. … etc. … etc. …" "Oh … yes," they replied. So the girls fixed about six of us up with dance partners for the evening, and we all went to the country club and had a very good time dancing and socializing.' With a naughty laugh, he adds, 'It was quite amusing, because, of course, having been cooped up on the ship for so long with no female company, we were hoping for some kissing and cuddling with these girls. However, we didn't get so much as a goodnight kiss! So there was a bit of moaning and groaning from the boys on the way back to the ship.'

Soon the boys, used as they were to being at the controls, experienced very unfamiliar territory. Left to the mercy of the powerful force of the sea, they were tossed about by gale force winds, terrific storms and the heaving gigantic swells of the sea. This made manning the anti-aircraft machine gun, which was housed in the top of the dummy funnel way above the ship's deck, a terrifying experience. Dad explains: 'There was a catwalk around the inside of the funnel and it was bloody high off the ship, and when we got into the huge swells the ship would roll over what seemed to be about 60 degrees and one would be standing straight over the water and it wasn't funny at all. Carey, who was with me, describes it as, "one moment the deck was beneath us, and the next there was nothing but a boiling sea below us." Anyway, we did our hours of duty manning Vickers gas-operated machine guns, all wrapped up in greatcoats and scarves, because it was bloody cold. We saw nothing, no submarines, kept up the zigzagging, and finally sailed into the Clyde River estuary towards Glasgow at 8.30 am on 6 December 1941. It was still pitch dark and absolutely freezing cold, and I thought, hell, this is not much fun. However, as the journey progressed and light started dawning, things warmed up with the welcome we received while sailing up the river.'

In *Heck What a Life!*, Carey Heydenrych writes:

> As we proceeded unhurriedly and majestically down the river, we were encouraged on our way by a tumultuous welcome from

THE BIG, WIDE WORLD

> all the men working on the shipyards and from the stationary ships standing-by along our path. They treated us to a magnificent display of great showers of water from the fire tenders and to a continuous cacophony of raucous noise from the sirens of the numerous ships lining our route down to the berth in Glasgow harbour. It was an unforgettably moving spontaneous reception we were given by those warm-hearted Scottish shipping folk.'

The ship eventually docked at Glasgow harbour, not Greenock, as planned, and the South African pilots disembarked. By 3.30 pm they had arrived at Glasgow Central Station. It was dark again and in blackout. Carey writes: 'The station platforms were scarcely lit by tiny blue lights, which did little to light the place.'

Dad's memory of the station experience centred around the unusual red tabs that the South African pilots wore on their epaulets, which all Air Force volunteers wore outside South African borders. He remembers: 'We were battling with our kitbags in the dark when three MPs (military police), who had mistaken us for general officers because of our red tabs, came charging up to us, stamped their feet, saluted smartly and said, "Can we help you, sirs?" We thought this was a helluva laugh because here we were, mere second lieutenants, and they were sergeant MPs, who were tough cookies, and here they were helping us with our luggage.'

> Ted Strever and I stood waiting on the platform for our train to pull in and, while waiting, we got talking to two Scottish WAAFs (Women's Auxiliary Air Force). They were intrigued by the red tabs on our shoulder straps that, we explained, we wore because we were South Africans. They [also] wanted to get a better view of us to see if we were white, so we moved under a lamp to let them get a closer look at us.
>
> Carey Heydenrych, from *Heck What a Life!*

The midnight express train trip to London was, according to Dad, 'freezing cold, no sleepers, just seats. With chattering teeth we had to keep stamping up and down the corridor to keep the blood flowing. And on the morning of 7 December 1941, we pulled into Waterloo Station, London, which was lined with posters reading "Japanese Attack Pearl Harbor", which we knew would mean America's entry into the war.'

Chapter 12

Puttin' on the Ritz

From Waterloo Station the young pilots boarded a train for Bournemouth, on the southern coast of England, just a hop across from the enemy coast of occupied France. Here they were to report to the Commonwealth Airmen Reception Depot. The trip gave them their first visual experience of the reality of the war.

> From the train window, as we left Waterloo Station, we saw mile after mile of the devastation wreaked on the East End of London by the Luftwaffe bombers. As far as the eye could see, along both sides of the railway track, were row upon row of working-class, double-storey tenement houses, which had been bombed out. Rubble was strewn all around the streets. The Germans aimed at the railway tracks but only succeeded, it seemed, in hitting the houses in the vicinity.
>
> Carey Heydenrych, from *Heck What a Life!*

Situated so near to France, Bournemouth's beaches were strewn with barbed wire. To create the illusion of a strong naval gun defence, blocks of concrete with pipes (imitation guns) cast into them lined the cliffs above the beaches. The elegant old wooden pier was now also rendered useless by having its centre blasted away. On arrival, the pilots were told that they would not be posted until January, and in the meantime they were to be accommodated in a hotel. This meant that they were to spend Christmas and New Year in Bournemouth with not much to do. The strict rationing of food, clothing and petrol was a constant reminder that England was an island suffering from the devastation of war, and the hotel's breakfasts, lunches and suppers left the boys longing for the hearty meals of South Africa.

PUTTIN' ON THE RITZ

> After breakfast we fell in for parade, in the street outside the hotel, for roll call and instructions for the day. Upon being dismissed ... we dispersed in small groups and rapidly did the rounds of the local cafés, ordering sardines on toast and more coffee to supplement our meagre breakfast.
>
> Carey Heydenrych, from Heck What a Life!

Dad says, 'In those days it was popular for restaurants to have tea dances where you could ask a girl to dance with you. Soon after we arrived in Bournemouth, Carey and I went to one of these restaurants. There were two girls sitting on their own there so we asked them to dance. After dancing with them and chatting to them we found out that they were sisters, by the names of Joy and Wendy Jones. Carey paired up with Joy and I with Wendy. This meeting started a long friendship between us all and they soon invited us to their home in New Milton, a little village about 10 miles up the coast from Bournemouth. Here we met their mother, who we nicknamed Ouma Jones, because she looked a little like Ouma Smuts, and their father, who for some unknown reason we nicknamed 'Old Pharaoh' Jones. They had a lovely small double-storey cottage called *Machrie*, and this became a home from home for us. Ouma Jones ran the Hovis bakery in New Milton, and upstairs from the bakery they had a room where Carey and I stayed on our visits to them. They were absolutely marvellous to us.

'One day, a notice went up in our hotel to say there would be a New Year's Eve dance. We thought this would be a nice way to repay the girls for being so good to us. We duly invited them, and damn-it-all if a few days before the event it was cancelled! So Carey and I trotted out to New Milton to tell the girls. When we arrived, before we could tell them, Ouma Jones started telling us all about the dresses the girls had managed to get under difficult circumstances. We suddenly realized we were in big trouble because the girls had used all their clothing ration coupons to buy their dresses. Now we were in a cleft stick and dared not say it was all off, so we decided that if we couldn't go to an event in Bournemouth we would go up to London, with Ouma Jones as chaperone. Ouma, who was always game for some fun, readily agreed to our proposal. When we contacted the South African Embassy and asked them to book us into a hotel and a dance they weren't very impressed, as everywhere was fully booked. Anyway, they eventually booked us into the Half Moon Hotel in Green Park, but they were unable to find us a dance. We decided that in London at New Year we were bound to find somewhere to dance. So we took the train to London, settled into our hotel, and called a taxi.

'I said to the taxi driver: "Please, you've got to find us a dance to go to." So off he goes and dumps us outside a rather posh-looking hotel, where a resplendent doorman gives us a suspicious glance as he ushers us in. Now, we were completely naïve, two green unworldly South Africans. Once in the place we were approached by a supercilious maître d' who looked at us down his nose and asked, "Have you a booking?" I replied, "No, we're terribly sorry ..." and before I could explain he said, "We are full." His snooty attitude got my back up so I said, "Look here, we are from South Africa and have come thousands of miles to help fight your bloody war. This is disgusting. We have brought these young ladies out, we have nowhere to go. If this is your English hospitality we think it stinks and you can go and jump in the lake." Rather surprised at this outburst he said softly, "Oh! Hold on a moment." He went off, and it was only then that we realized we were at the bloody Ritz.

'He returned to lead us through the dining hall, which to our dismay was filled with admirals, generals, vice marshals, air vice marshals, commanders and aristocrats. Here we were, two lowly second lieutenants in army uniforms with wings on them, having some funny looks thrown at us.' Dad laughs, and goes on: 'Talk about feeling out of place! Anyhow, with our rapidly shrinking bravado we settled at a table and ordered drinks. As we perused the menu, Carey and I looked at each other in horror. A dinner was two pounds ten shillings. My gosh! It was a bloody fortune in those days. We counted up all our money, and we had twelve pounds between us, which allowed for one more drink and dinner for all of us. The girls thought it was all hilarious, which put the spirit back into our little party. We ended up really having a ball, dancing amongst the top brass, and got many a funny look from them. As the Irving Berlin song goes, we were really puttin' on the Ritz.'

> *If you're blue and you don't know where to go to*
> *Why don't you go where fashion sits*
> *Puttin' on the Ritz ...*

'So comes the end of the evening,' Dad continues, chuckling, 'and we give the waiter all our money to pay the bill. He returns with sixpence change and says, "Your change, Sir." We say, "No, no that's all right." He looks

PUTTIN' ON THE RITZ

disdainfully at us and again says, "Your change, Sir!" making it quite clear that he wasn't going to take our tickey-for-a-church-collection sort of money. When we got out of the hotel we saw a newspaper van and explained to the driver that we had no more money and asked him to take us to the hotel, which he did. Back at the hotel we continued the party with Ouma Jones, and there was much hilarity over ending up at the Ritz. At least the girls had a wonderful time and will probably never forget that evening. Next morning we returned to Bournemouth. It was New Year's Day of 1942.'

Chapter 13

Dive-Bombing and Boyish Pranks

> *There is a kind of character in thy life,*
> *That to the observer doth thy history*
> *Fully unfold.*
> William Shakespeare, Macbeth

On 26 January, the South African pilots were posted to the operation training unit for the torpedo bomber aircrews at Chivenor, an airbase between Barnstaple and Ilfracombe in North Devon. Some of the pilots who had not opted for torpedo bombers were sent to other training units, and so, yet again, close friends were parted. Carey Heydenrych,[1] Dad's good friend, was posted to Thornaby for training on Hudson bombers. Neither realized when they said their goodbyes that both would be shot down at some stage, taken as POWs, and would participate in what later became riveting escape stories.

Among those who were to be on torpedo bombers with Dad were Don Tilley, Gill Catton,[2] Hugh Sheldon and Wally Morgan. When they arrived at the new training base ...

The first thing they expected us to do was to go out on parades for physical training in vests ... and, my God, it was bloody freezing, being January. Well, we demanded to see the course captain, quite a good chap actually, a Squadron Leader Hartley, I think, and we said, 'Sir, let's just be reasonable about this. We're from South Africa, where it is hot, and our blood is thin. You're going to kill us if you put us out in this cold weather.' He looked at us ... and he was a bit of an old soldier ... and he thought for a while, and then he said, 'Ah, I can understand it is very cold for you South Africans, so you can do PT in the gym.' So we didn't get out of doing PT after all!

DIVE-BOMBING AND BOYISH PRANKS

At Chivenor, the pilots were at last going to train in wartime aeroplanes – an exciting prospect for Dad because of what he had read about the Beauforts.

I read in the aeroplane magazine *Public Knowledge* that the Beaufort was the fastest torpedo bomber at the time, and capable of flying at 300 knots and blah blah, so I was really keen to try one. Well, I got an instructor, and I won't mention his name, because he was the idlest bugger I've ever come across. He also had a thick dirty ring around his neck and I thought, 'Jesus, don't these bloody Englishmen ever wash?' Anyway, I got fed up when everybody else was getting their chance to fly solo in the Beauforts and I hadn't even got into one because he was always late and messing about. But eventually I went solo on the Beauforts, and at Chivenor we practised dive-bombing on a big raft that was anchored in the nearby bay. This was quite exciting really, and great fun. On one occasion, I came in low, watching the target, judging it, with my nose down in a fairly steep dive … and I was so intrigued with this that when I let the dummy bombs go and started to pull out I had left it so late that I could see the blinking joints between the wooden planks of the raft. I must have pulled out at about between 6 and 10 feet above the water, and I thought I was going to hit it! That was quite scary, but a good lesson, because one realizes that you get so focused on the task you are doing you start ignoring the aftermath or pullout.

As trainees we also had to do what was called 'Paraffin Pete's job'. Paraffin Pete was the nickname given to the duty pilot who manned the chance light, a huge searchlight on the runway that we operated as aircraft came in to land at night – all out of date today, of course! In addition to the chance light the duty pilot had two Aldis signal lamps to operate, one red and one green. The red light when flashed signalled to an incoming pilot to remain airborne, and the green signalled the all-clear for landing. Those original RAF aerodromes had runways in the shape of a triangle so that the aircraft landing and taking off were never far out of wind.

Now, Chivenor had two runways, one going up towards the hill, and the main one, which ran parallel to the river. One particularly busy night, aircraft that had been out on navigation exercises started to return and I was on duty out on the runway.

There was no radio contact with the pilots and so all the controlling was done visually. I picked up the first aircraft in the circuit, he flashed his light to land, and I flashed the green, giving him the all-clear. Then another one arrived and flashed for landing and I flashed red, because there was a bloke in front of him. Well, this was all going fine and I was getting them down, until one Beaufort came in and did a terrible landing. He bounced to high heaven, revved up and went round again. I thought, 'Well, he'll be a bit upset so I'll keep an eye on him and won't hold him up.' As soon as I could, I gave him the green and down he came. On his approach, I put the chance light on and – Oh God! His under-cart was up and it was too late to give him the red. Well! A magnificent firework display erupted as his belly hit the runway, the sparks were like enormous Roman candles on Guy Fawkes Day. Fortunately, the aircraft didn't catch fire, but unfortunately came to rest in the middle of the bloody runway, while I still had five aircraft up in the air waiting to land.

So, in haste, the chance light, Aldis lamps and all the goose-neck flares[3] that line the runway had to be moved across to the other runway. The secondary runway ran towards the hill and ended at right angles to a thick 6-foot high hedge separating the runway from the railway line on the other side. Anyway, I brought in one Beaufort and then another and then another, and one bloody fool taxis around and around and prangs right into the Beaufort that had crash-landed on the main runway. What a palaver!

Then a Hudson comes in, approaching from beyond the hedge that ran across the approach. I'm watching this bloke, ready to put the chance light on and all of a sudden his lights disappear behind the hedge. This aeroplane then crashes through the hedge and touches down before the runway – by which time I had put the chance light on, but he bounces all over the place and swings off towards some parked Beauforts, and I thought, 'Oh God, here's another one that's going to be written off.' Thank goodness he stopped just before hitting anything. Well, I was pleased when that night was over and, I promise you, I never did Paraffin Pete duty again. I thought it was far too arduous, far too dangerous, and far too responsible. ... I didn't want to know about it after that.

In the middle of the course, the trainees were given a forty-eight-hour pass so Dad hopped on the train and went to see the Joneses in New Milton for the weekend. When he arrived he got a lovely surprise – old Pharaoh Jones had managed to find an old 1929 Austin 7 Sedan that was for sale. So on the Sunday, he bought it for £10 for Dad to use while he was training, and would sell it again when he was posted elsewhere. They managed to scratch a couple of petrol coupons together so that Dad could drive it back to Chivenor. Dad says ...

Well, this was bloody hilarious. Talk about going it blind! Firstly, I wasn't used to driving in England. Secondly, all the road signs had been taken down in case of a German invasion. And thirdly, there was blackout, and car headlights were covered with a flap, like a letter slot flap, so enemy aircraft couldn't see

DIVE-BOMBING AND BOYISH PRANKS

them from above. This resulted in minimal visibility for the driver. So, in youthful ignorance, I set off at three o'clock thinking I would make it by the time darkness fell, because it was only dark at about seven o'clock. I had written down all the names of the places I had to drive through: Exeter, Taunton and along to Barnstaple, which was the village at Chivenor.

Well, without signposts I had to keep stopping to ask for directions and was getting very suspicious looks because of the unfamiliar uniform I was wearing. So, in this old car and with all the stops, it started to get dark just outside of Exeter. Thank God there was no traffic on the roads. I ended up going through Barnstaple in the pitch dark and could see bugger all. I had to follow my nose and drive along the edge of the pavements to stay on the road. Instead of arriving at Chivenor at seven, I arrived at ten; the trip had taken seven hours instead of four. However, all the boys were delighted, for we now had a car to get around in during the second phase of our course.

We had such a lot of fun in that old Austin, which we called *Stooge*. It was only designed as a two-seater, but six of us would pile into it and old *Stooge* took it. One night we decided to go to a pub in Braunton for a drink … and

Carey Heydenrych changing a puncture on Stooge – with many little onlookers – during a trip from Bournemouth to Southampton. Author's collection

ON LAUGHTER-SILVERED WINGS

Mr Tilley, Don ... he's a bugger, you know ... when we got into Braunton, because of the blackout I parked in the main road. Later, after having quite a few drinks, we emerged from the pub, and there was this gentleman in a dark blue uniform next to *Stooge* and as we approached he said, 'Good evening, Sir.' I said, 'Good evening, officer.' He asked, 'Is this your car, Sir?' I replied, 'Yes, it is.' He said, 'Sir, you are in a no-parking area.' I say innocently, 'Oh, am I?' And what follows is a classic example of Don Tilley's perverted sense of humour. He pipes up and says, 'Officer, I told this man that he was in a no-parking area and he told me to do some very rude things ... very rude, and he wouldn't listen to me. I'm a very law-abiding citizen and I was absolutely horrified at his irresponsible behaviour.' The other blokes are killing themselves with laughter because he was getting me into such trouble. At that time I only had a South African driver's licence, which I didn't have with me, and hadn't obtained an international licence for England. However, the officer gave me a warning ticket but insisted I go and buy a licence. So off I went to the police station to explain myself. It cost me a quid, which I put in the orphan and widows' fund collection box. Tilley thought it was a helluva joke.

On another occasion, we were all instructed to do a 6-mile run through Braunton and back to the base. The powers that be didn't trust our South African group, so they organized for a flight sergeant to be at the midpoint to stamp our arms with a rubber stamp as proof that we were there. Wally Morgan had for some reason been let off doing the run, so we worked a plan with him. He was to take the Austin and pick us up as soon as we had run out of sight of the airbase. He was then to drive us to Braunton, where we would have a few beers, then drive us near to where the flight sergeant was waiting and drop us off. We would have our arms stamped and run on. As soon as we were out of sight of the flight sergeant he would pick us up, we would go for a few more beers, then he would drive us back to near the airbase and we would run back in through the gate.

Well, this was the plan, except that the flight sergeant had hidden down a side street and we drove straight past him, all hanging onto this motor car and in full view of him. We all shouted, 'Wally, stop, for God's sake,' and we all jumped out and he drove out of sight again. The flight sergeant looked at us with a twinkle in his eye, stamped all our arms. Off we go and jump back in the car again. When we eventually pitched up at base camp running like sprinters, without a drop of sweat on us, the blokes in charge just shook their heads. Thanks to the flight sergeant not a word was said. The RAF was bloody good to us, we got on well with the senior officers and there was fantastic camaraderie amongst us.

When we had finished the course, each pilot was given a crew, and we were posted to Abbotsinch, near Glasgow, to do a five-week course of proper

DIVE-BOMBING AND BOYISH PRANKS

Ted on the train to Abbotsinch, 1942. Author's collection

torpedo training. Before leaving for Abbotsinch, we were given a few days' leave. Carey and I went to say our thanks and farewells to the Joneses, and, sadly, to return *Stooge*.

ON LAUGHTER-SILVERED WINGS

We arrived at Abbotsinch in spring. It was mid-April, and the weather was particularly good. With long days and short nights, the sun only set at about ten-thirty and started to rise again at about two-thirty. Because of this, the instructors pushed us to get as much flying practice in as possible. Now, torpedo work is low-flying work, and low flying is normally taboo in the Air Force because so many blokes who had attempted it had written themselves off. So, the focus and extent of our training on this course was to get us thoroughly trained in low-flying techniques. We used to climb to torpedo height at about 80 feet, drop the nose and whizz over the water, with the props about 3 to 5 feet above it, and then pull out.

From left: Morse Currie, Wally Morgan, Ted, and Don Tilley in Abbotsinch, 1942.
Author's collection

DIVE-BOMBING AND BOYISH PRANKS

From left: Morse Currie, Don Tilley and Ted with a torpedo, Abbotsinch, 1942.
Author's collection

These exercises were carried out in an aircraft called the Blackburn Botha, named after General Louis Botha. These, ironically, although designed as torpedo bombers, wouldn't fly with a torpedo, so the exercise drops were carried out without a torpedo, but with a camera gun that was mounted on the target ship, which ran up and down the coast for our practice purposes. The camera gun would assess our deflection and height as we came in for a dummy torpedo drop and it would signal these, in Morse code, back to Command.

Anyway, one day I was flying on an exercise and my wireless operator picked up a signal going back to base. It read 'Botha number four,' which was my aircraft, 'is flying dangerously low,' and, with a squirming stomach, I thought, 'Uh-oh, I'll be in for it from the CO when I get back.'

Sure enough, we landed to a message: 'Skipper of Botha number four report to CO immediately.' So, nervously, I go into the office of our CO, a Squadron Leader Darling. And, hell! He *was* a darling. He was a helluva nice chap. He says, 'Ah! Strever.' I said, 'Yes, Sir.' He says, 'I got a message to say that you were flying dangerously low?' I replied, 'Uh ... yes, Sir.' He smiles and says, 'Well, it's about time some of these chaps started learning to fly torpedo bombers!' and continues, 'Well done, keep it up, but don't hit the bloody water.' And I, hiding my relief, said seriously, 'Yes, Sir. Thank you, Sir.' Off I went, as chuffed as hell.

On our off times we went into Paisley, a town about 12 miles from Glasgow. One day, to celebrate the completion of our flying training, we all went into Paisley, to a rather posh restaurant called Regano. All in a festive mood and with a few drinks under our belts, we started singing. We were promptly chucked out of the restaurant by the doorman, who stated, 'An donna yee cum back heere. Yee remember wee donna waunt yer kaind heere.' It seemed no one appreciated our slightly bawdy sing-song. It just wasn't done in Regano.

Another memorable place we frequented, once we had finished our training, was the fantastic Palais de Danse in Glasgow. It was huge, absolutely huge. There was a revolving stage in the centre of the hall on which the bands would play. When one band had their rest, the stage would revolve and another band would continue, so there was continuous music and dancing. There was never a dull moment in the Palais de Danse as it tended to be a bit rough, but we had a very good time there.

Having finished our course in less than five weeks, we now had time on our hands, so to keep us busy it was decided we must go on a route march. To make sure we would actually do it, they put us in a bus and drove us all to the other side of Paisley, where we would begin our march. On the way, Don Tilley and I and two others agreed that it would be much more enjoyable to play snooker in Paisley. When the bus stopped, and all the blokes were disembarking, we lay on the floor. When the last chap got off the bus the driver looked around, and as he saw us we put our fingers to our lips. He shrugged, and drove off. We all said, 'Thanks old chap, just drop us in Paisley.' This he did, and as we alighted we thanked him again and promised, 'You'll be rewarded in heaven for this.' Well, we had this all taped, or so we thought. We would have a few games of snooker, judge the time it would take for the other blokes to march back, catch a tram to Abbotsinch, jump off a few stops from camp, hide in a hedge until the other blokes came past, and then join the back of the line as they marched into base. But it didn't quite work out like that because we miscalculated the time, and when we got off the tram the others were already back in camp. So there we sat, like bloody clots, in the hedge. We were stuck with our fate and had to face it. When we arrived at the gate we were caught fair and square. The next thing is we are up in front of the CO, dear old Squadron Leader Darling. He looks at us in dismay and says, 'I really don't know what I am going to do with you chaps ... uh ... I

DIVE-BOMBING AND BOYISH PRANKS

think that you had better go on a cross-country run on Saturday with Squadron Leader Larry Gain.'

This didn't faze us, as we didn't have a very high opinion of this Squadron Leader Gain. He was a tall thin chap and we called him, rather irreverently, Dan-Dan-the-Stratosphere-man, because on torpedo exercises he would fly at about 30 feet, which we all thought was a bit high.

Anyway, we said, 'OK, Sir, thank you, Sir.' We then trotted off to the mess and the other blokes asked what had happened. We reply, quite nonplussed and laughing, 'Oh, it's no problem, we're going on a cross-country run with Larry Gain.' Then they all started laughing, and said, 'Do you know who Larry Gain is?' We said, 'What do you mean?' They replied, still laughing, 'He's the RAF cross-country champion, you fools.' Well, levelled by this news, we rushed back to the CO and said, 'Sir, by the time Squadron Leader Gain is finished with us, we're going to have sprained ankles, sore knees, blisters on our feet, and we'll be full of aches and pains. We really don't think this is a very good idea.' He looked at us with a twinkle in his eye and said, 'You bad buggers. I'll cancel the whole thing. Now – get out of my office!'

When we were finished at Abbotsinch we were posted to 217 Squadron at Leuchars, which is an aerodrome near St Andrews Golf Course. When we arrived, we were told that the squadron had already been posted to the Far East, Ceylon, and that we would be flying out to join them. First, we had to have all the necessary injections to guard us against tropical diseases, one of which was yellow fever. You know, this posting to the Far East was a typical example of trying to lock the stable door after the horse had bolted. It was now mid-May 1942, and at Easter 1942, the Japanese had attacked Ceylon, at Trincomalee Harbour. Ceylon, at the time, was quite unprepared for the onslaught, with minimal defence systems in place. In fact, tragically, two of my friends were killed during the attack. One was Derrick Knight, who before the war had been a fantastic rugby player and had played wing for the

A Bristol Beaufort Mark 1 on the runway at Leuchars training airbase, Scotland, 1942. Author's collection

ON LAUGHTER-SILVERED WINGS

A Beaufort training exercise over Scottish countryside (picture Taken from Ted's aircraft), 1942. Author's collection

Transvaal team. The other was MacGlashon. Both chaps had trained with me in Number Three B course at Induna, in Rhodesia. They were flying Blenheim Bombers for 11 Squadron and were sent out to attack the Japanese ships. And, I'm telling you, they attacked the ships! The Japanese fighter aircraft shot down many of our blokes while they were attacking the ships. It was an absolute debacle, and I don't think many came back from that raid.

So this is where we were being sent – after the fact! However, after having all our injections we were given ten days' leave. We visited St Andrews Golf Club, and sat on the veranda in the late afternoon having a beer. It was the end of May, which is a lovely time in Britain. The evenings were beautiful and twilight lasted until about ten o'clock. The spring weather was warm and mild, all the flowers were blooming and the trees were in leaf. Ah … it really was a splendid sight.

After our leave, my crew and I were posted to Lyneham, in the south of England. My crew consisted of Pilot Officer Bill Dunsmore as navigator, Sergeant Dick Ellis as wireless operator, and Sergeant Bob Gray as air gunner.

DIVE-BOMBING AND BOYISH PRANKS

When we arrived at Lyneham, we were sent down to the Bristol Aircraft Company at Filton to collect a brand new Beaufort. We were surprised to be ferried to Filton in an Anson piloted by a woman ferry pilot. The fact that she was a woman didn't worry me at all – but it was just so unusual. By the way, she was a magnificent pilot and deposited us safely at Filton. There we were given this new Beaufort. I tell you, when I took this thing off the ground at Filton Aerodrome it was one of the worst Beauforts I had ever flown. It flew sideways, like a crab. I couldn't trim it out; it was terrible. When I got it down at Lyneham I had a helluva moan about it and they said, 'They're all like that.' I objected, 'No, we have got to do something about it.' Well, something was done and we got it to fly a bit straighter.

All the Beauforts going out on long-distance flights had to have fuel consumption tests. As we were going to fly to the Far East we would be flying from the United Kingdom to Gibraltar, from there to Malta, from Malta to Egypt, from there to the Middle East, from there to India and from India down to Ceylon. All these legs of the journey were on average seven to seven and a half hours each. Normally, the Beauforts' average fuel range was about three and a half hours.

Once the long-range tanks were fitted and filled with fuel we set off to do a consumption test. We flew from Lyneham in the south of England, across central England, onward up to the north, over Scotland and the Hebrides, and then turned out over the Atlantic just north of Ireland. We had to stooge around way out over the ocean for a few hours and then fly all the way back to Lyneham. When we returned we had flown for eight hours on the fuel. It was the longest flight I had done.

For our Far East posting, we were now briefed to fly from Lyneham down to Portreath in Cornwall. On leaving Lyneham we were loaded with spares and supplies for the Middle East and Far East, and all our kit. On the night of our arrival at Portreath we were briefed for the following day. We were told that we would possibly see ships anchored off Lisbon in Portugal, and instructed not to fly over Spain, and to stay out of sight of land. The aircraft were refuelled and ready for take-off the following day.

Next morning we sat at the end of the runway at Portreath Aerodrome, which has a runway facing out to sea and a 400-foot cliff at the end of it that dropped away to the sea below. We were loaded to the hilt with full fuel tanks, full of spares and supplies, and all our kit and clobber. We were definitely above the normal maximum gross weight. Anyway, off we tootled, roaring down this runway, we were heavy on our wheels and just managed to stagger into the air, drop down over the cliff, losing about 150 feet, and slowly building up a reasonable flying speed. We flew over the Scilly Isles and, with all the weight we had, were flying at about 110 knots, just staggering through the air in a nose-up attitude. There was nothing extraordinary about the trip except that the plane didn't fly very well, and the 300 knots capacity that I had read about was still grating me. It got better once we had burnt up some fuel and we flew over the Bay of Biscay. We landed at Gibraltar at about three o'clock in the afternoon. Once I had parked the aircraft we were taken to the mess.

While having a drink in the mess we met the crew of a British Imperial Airways flying boat. The captain of this flying boat was intrigued that we had just arrived in a twin-engined Beaufort from England and he asked me, 'How many flying hours have you got?' 'Ah …' I replied proudly, 'I've got 190 flying hours.' 'Is that all?' he exclaimed. 'And they sent you from UK to Gibraltar with 190 hours? Taken aback, I said, '*Ja*, what's wrong with that?' He just shook his head in astonishment.

My crew and I were then briefed for the following day's flight to Malta. I went to bed pleased that one leg of the long journey to Ceylon was behind us, and fell into a peaceful sleep.

While they slept that night, Dad and his crew were blissfully unaware that in a few days they would be co-opted into the midst of a terrifying battle with eleven war ships of the Italian Fleet. This, their first encounter with the enemy, they would have to face alone. For them all, the horrible reality of war was just around the corner. They were about to go on their first operation, which is where my story began. Unbeknown to them, their planned destination, Ceylon, was still five months away.

Chapter 14

Where There's a Will, There's a Way!

*In all acts of creativity and initiative,
The moment you fully commit yourself Providence moves too.*
William Hutchinson Murray, *The Scottish Himalayan Expedition*

*Do not fear the Laistrygonians, and the Cyclops, and the angry
Poseidon
You will never meet such as these on your path
If your thoughts remain lofty
And a fine emotion touches your body and your spirit. ...*
C.P. Cavafy, from the poem *Ithaca*

My crew consisted of Pilot Officer Bill Dunsmore RAF as navigator, the oldest of the crew, a retiring, rather cautious British fellow; Sergeant Ray Brown as wireless operator, a strong and stocky New Zealand farmer; and Sergeant John Wilkinson as air gunner, a tough, thickset chap who was also a New Zealand farmer. These two New Zealanders of the Royal New Zealand Air Force were affectionately known as 'Brownie' and 'Wilkie' respectively.

We were preparing for take-off from Luqa Aerodrome, Malta, in the midst of an air raid, when suddenly one of the Spitfire boys descended from the fray above, swerving dangerously from side to side. Obviously wounded, and unable to control his Spitfire, he hit one of the stone walls that surrounded the dispersal area of the aerodrome, the wing of his aircraft spinning into the air. This fatal crash happened within yards of us. The horror didn't do much for our morale at this point in time.

Earlier that day, 28 July 1942, at seven o'clock, Spitfire pilot H.C. Coldbeck had sighted two Italian destroyers escorting a merchant vessel and travelling towards Africa, along the west coast of Greece. This he had reported, on his return, to headquarters in Malta. We were duly briefed on the situation and, at nine o'clock, nine pilots and crews were gearing themselves up for another torpedo strike on Rommel's supply ships.

> The vessels spotted by Coldbeck were the armed transport *Monviso* ... carrying a mixed cargo of war supplies. She was escorted by the Italian destroyer *Freccia*, armed with four 120mm guns and four 37mm flak guns. The other escort was the modern torpedo boat *Caliope*, armed with three 100mm guns as well as six 37mm and two 13mm flak guns. These vessels could put up a very powerful defence against attacking aircraft. They had set out the day before from Brindisi and their destination was Benghazi.
>
> Roy C. Nesbit, Reported Missing

In spite of the unnerving experience of the Spitfire crash we took off, formed up in loose formation and set course flying eastwards at a low altitude of 50 to 100 feet. We were led by our commanding officer, Squadron Leader Pat

Wilkie and Brown in the cramped fuselage of the Beaufort, Malta, 1942. Courtesy of the Roy C. Nesbit collection

WHERE THERE'S A WILL, THERE'S A WAY!

Gibbs, who was the most incredible squadron commander I had ever come across. He was utterly fearless. I was in Beaufort one, serial number L9820, with my regular crew, and I was to fly number two to the CO, flying on his right side.

In addition to it being a sweltering day, especially at low level, my bloody heater in the cockpit was jammed in the ON position. I was absolutely roasting, so Dunsmore, in an attempt to cool me down, used his clipboard to deflect cool air onto me from the storm window. Our Beaufort Mark 1s had been what we referred to as 'tropicalized' – which meant there was a bloody huge air filter sticking out of the top of the motor, and therefore performance in the heat of a Mediterranean summer, compared with the cold weather in England, was laboured. They were supposed to cruise at 140 knots, but on these strikes at 50 feet above the sea, we just managed to get to 115 knots. So we were not on the 'step', like a boat not quite on the plane. It was ghastly! They were very demanding to fly in these conditions, and required constant control and concentration. Soon, we became drowsy and lethargic on the three-hour flight in the hot, heavy and lumbering aircraft. On approaching the target area, however, we snapped alert, with adrenalin now pumping through our veins in the nervous anticipation of impending action.

At 12.15 pm, the enemy coastline appeared on the horizon and we dropped to 20 feet above the sea. Five minutes later, we sighted the target, a merchantman escorted by two destroyers about 5 miles off the Greek coast. We approached from astern, and the ship immediately started evasive action by turning to port. Our CO countered, by turning our formation into an attacking position. At 1,500 yards, and still only 20 feet above the water, the flak started bursting around us. Unusually, no tracer was noticeable in this strike, only flak, which is just as dangerous, but not half as frightening as having tracer flashing around us. At approximately 1,000 yards from the ship, with the CO ahead of me to my left, I climbed to about 80 feet, accelerated and steadied the aircraft at 140 knots and with a perfect attacking position, slightly off the starboard bow of the ship, I released, and shouted to my crew, 'Torpedo dropped!' and noticed to my satisfaction that the ship kept turning into the path of my torpedo.

At this stage, the CO in front was flying towards the bow of the ship, which he flew straight over from bow to stern. I reasoned that it would be safer to fly below deck level on the ship's port side, so I quickly made a slight turn to starboard, dropped down to a few feet above the water and, under enemy fire, flew down the length of the ship – when I was about mid-ship the aircraft gave a sudden violent shudder. 'You're so dammed close,' shouted Brown, 'I can see the deck crew's uniform buttons shining in the sun.'

Once clear of the target, with loud banging noises and violent vibrations coming from our engines, I looked out and saw smoke pouring out of our port engine. Dunsmore, who had been in the nose, got up next to me, looked at the starboard engine and shouted, 'Christ, this one's had it as well!' Both engines were streaming smoke. As I still had power, I formed up with the CO. Almost immediately, the oil pressure failed on the starboard engine. I realized we were not going to make the two and a half-hour trip back to Malta. I waggled my wings in farewell to the CO, climbed as fast as I could, and, with no alternative, turned apprehensively towards the enemy coast. I did this to be as close as possible to land, if and when the plane went down.

The CO remembers:

> I saw immediately that the right-hand aircraft was in trouble; white smoke pouring out in a trail behind the starboard engine, and after a few minutes the pilot broke formation and turned back towards the enemy coast. ... The loss of Ted Strever, the pilot of my right-hand aircraft, was a severe blow, for he had followed me with some skill on every one of the previous attacks and was about to become a section leader.
>
> Wing Commander Patrick Gibbs, DSO, DFC and bar, *Torpedo Leader*

Within seconds, the starboard engine seized, and the port engine began to vibrate vigorously. Now, at 300 feet above the Mediterranean, with the deafening noise around me, I shouted the order to my crew, 'Ditching stations!' I had no altitude to play with, and unhappily had to ditch downwind, which, of course, increases the speed. I automatically applied landing procedure – trim, throttle back flaps only half down – and the surface of the sea, a dark blue-green wall, came rushing towards us.

Going at approximately 110 knots, the tail wheel caught the water and I yelled, 'Ditching now!' with the stick well back. And we hit the sea with a gut-wrenching, whip-lashing jolt and shudder, the open doors and empty cavity of the bomb bay in our underbelly caught, cupped and dug into the sea, ripping the aircraft in half. Dunsmore had shot past me like a bullet, slashing his arm in the process. The impact threw my head forward so violently that my leather flying cap and earphones flew off. As the aircraft tore apart, I felt the nose dive down and a huge green wave of water smashed into the cockpit over my head, enveloping Dunsmore and me in sudden liquid green silence.

Totally underwater and sinking, with the nose tilting at a 70-degree angle in the sea, now panic-stricken, I tried my damnedest to move quickly but, like a bad dream, every move was slow and laboured. I eventually wrestled free

WHERE THERE'S A WILL, THERE'S A WAY!

from my seat harness. Dunsmore, now above me, was standing on the step next to the spar, battling to open the forward escape hatch. I swam up under him, desperate for air. He opened the hatch, got his head out. By God! He just stood there catching his breath. I hammered him on the back of his knees with my elbow, trying to tell him, in splutters and bubbles, 'Get a bloody move on and get out!', which he promptly did, and I climbed out after him, gasping and gulping at the fresh air.

We crouched, drenched and bedraggled, hanging onto our wreck, to find Brownie and Wilkie sitting like two gentlemen in the dinghy, both bone-dry. They hadn't even got their feet wet! Anyway, we climbed aboard with them. The tail of the aircraft was floating about 30 yards away and no sooner had we pushed away when, with a soughing gurgle, the front of the aircraft disappeared. We were all alone, in a tiny open dinghy, floating in the sudden uncanny silence in the middle of the ocean. I immediately said to my crew, 'What do we have with us?'

Dunsmore replied, 'I've got my bag, so we have maps, charts and a Verey pistol.'

Just then we heard a drone, and an Italian Macchi MC.202 fighter emerged above us. As the enemy had been known to shoot up dinghies, I said to the blokes, 'If he makes any threatening moves, get over the side and dive as deep as you can.' Thankfully, he merely circled us and pushed off. We then tried to paddle towards land, but there was such a heavy drag on the dinghy that we had absolutely no success. (The reason, we only found out later, was that the anchor pockets underneath it were out and cupping the water.) Within an hour there was another aircraft approaching us, an Italian Cant Z.506B seaplane, which circled us and landed some 30 to 50 yards away. While it was circling and alighting we very promptly wrapped our maps and charts around the Verey pistol and threw it overboard, so as not to have anything incriminating on us. As paddling anywhere had proved useless, when the aircraft was stationary I dived overboard and swam to it. I was provided with a rope. I swam back to the dinghy, attached the rope to it, and pulled it to the aircraft.

When we were all up on the wing of the Cant, the Italian crew treated us with the utmost courtesy. After searching us, they gave us a cigarette each and ten minutes to stand there and smoke it. The Italian crew consisted of First Pilot Lieutenant Chifari, Second Pilot Lieutenant Saliermo and crew members Schisano, Mazzeri and Logozatolo. We were then ushered inside and because the sea was too rough to take off, the pilot taxied to calmer water and took off. After flying for two hours we landed at Prevesa, a port on the Greek coast, and alighted like VIPs onto a jetty crowded with onlookers who wanted to get a glimpse of the British prisoners. We were taken from there to the officers' quarters, given dry clothes and a slap-up meal of steak, eggs and tomatoes.

The British POWs and Italian crew standing on the wing of the Cant on arrival at the bay of Prevesa, 28 July 1942. Foreground, from left: Dunsmore, Wilkie, Chifari, Schisano, Saliermo, Logozatolo. Behind, from left: Mazzeri, Brown (bending down), and Ted (side view), looking very worried. Author's collection

After eating the bully-beef fare in Malta it was manna from heaven. We were treated as fellow fliers, and one Italian officer, who spoke English, was very disappointed that none of us played bridge; he thought that all Englishmen played the game.

After the meal we were subjected to a very mild interrogation, where the intelligence officer told us, 'Your main interrogation will take place on the mainland, and then you will really be prisoners of war.' They seemed to accept that we would only divulge our names, numbers and ranks. They also seemed to think we were operating out of Africa because of my South African Air Force uniform. We were issued with pens and paper to write letters to our families, which would be sent via the Red Cross. I wrote to my mother, 'I am safe, don't worry about me.'

That night we were given officers' rooms but were told that as we were under armed guard we would be shot if we tried to escape. However, not to be deterred, once alone in our rooms, Wilkie and I started planning an escape strategy. We thought that in the morning we would be put on a boat to the mainland and then on a train or truck, and transported overland through Yugoslavia. If this were to be the case, we would bunk off these vehicles and find our way on foot through the mountains to Turkey.

Well, the next morning they returned our dry clothes, and to my surprise,

WHERE THERE'S A WILL, THERE'S A WAY!

my little thumbnail compass was still secreted in a tiny hidden pocket in my shirt. I thought, 'This is a Godsend!' At breakfast, we were told we would be flown to Taranto, and my heart sank. I thought, 'We've had it, there's no chance of getting away now.'

Before they put us on board we were each given a couple of packets of cigarettes, and told to enjoy them because we would not get more where we were going. Being the skipper, I was ushered into the plane first, then Dunsmore, Brown and Wilkinson. It was a fairly large fuselage and Dunsmore was seated on my right, with Brown next to him and Wilkinson on my left, nearest to the Italian crew. The two pilots got in and sat one behind the other. Three other crew members climbed aboard, and they battened down the hatches. We had expected at least seven or eight crew members to be accompanying us, but only five alighted. The wireless operator, stationed at his desk, sat between us and the armed guard and the flight engineer sat up in the front with his feet over the bomb well step, facing away from us. With only five Italian crew members and four of us, things were looking up and I started considering the possibility of taking over the aircraft. Next to me was a Chianti bottle full of oil so I surreptitiously dragged it to my side, but the wireless operator saw me and took it away. Wilkie, thinking exactly the same, leaned over and whispered, 'Can you fly this thing?' '*Ja*,' I answered quietly, 'any pilot can fly a plane when it's in the air.'

An insistent '*Zitti*!' came from the guard, and we knew it meant keep quiet. I was thinking that the only odds against us was the guard's revolver, in its holster and strapped to his waist. I was confident that with this revolver in our possession we would be home and dry.

We taxied out and at 9.15 am we took off for Taranto in Italy and POW camp. Our ETA at Taranto, we had been told, was 11.00 am. We had ninety minutes' grace for a chance at freedom.

Wilkie and I knew that Brown was with us, and although Dunsmore was rather a cautious chap, once things were happening he would assist us. So, with our minds sharply focused and bodies tensed, we waited for the right moment to act. We knew that we had to get the wireless operator out of the way in order to get to the guard and, fortunately, Wilkie, a thickset and tough New Zealand farmer with big strong hands, sat nearest to him. After an hour of flying, Wilkie suddenly tapped the wireless operator on the shoulder and, pointing out of the back window, shouted 'Spitfires!' When the wireless operator swung his head up and around in fright, Wilkie gave him a haymaker on the jaw.

Alerted by Wilkie's first call of 'Spitfires!', I was ready to move the moment he swung back his fist to strike. As he connected, I jumped the wireless operator and shoved him back towards Brown and Dunsmore. Meanwhile, in an instant, Wilkie, quite incredibly, ripped the revolver, complete with holster and belt, from the guard's waist and threw it to me. Before the guard realized what was happening, I had the revolver out and pointing at him. He frantically patted his hip, screaming, '*Mia pistola, mia pistola*!' Wilkie and I pulled him behind us for Brown and Dunsmore to deal with, and I thought we had made it ... but suddenly, Wilkie screamed, 'Look out! He's got one!' And there in front of me, crouching in the bomb well, was the first pilot, Mastrodicasa, peering over the step with just his eyes showing and pointing a Beretta pistol at me. At this stage, Wilkie had his watchful eye on the second pilot, Chifari, who was flying the plane and was therefore of no immediate threat. For a brief stationary moment, the armed Italian pilot and I weighed up our options. I was suddenly very conscious of the large 6ft 2ins target area I presented, compared to my opponent's 4-inch forehead. He didn't want to shoot towards the back, where his crew were tied up, and I didn't want to shoot towards, and possibly damage, the aircraft motor, which was just behind him in the bomb well. Just at this moment, an uncanny bit of luck occurred in the form of the flight engineer. He had obviously been dozing in the front, and was quite unaware of the events behind him. Now awake, with his eyes fixed on the armed first pilot who he thought had gone mad, he started crawling backwards down the fuselage. With one accord, Wilkie and I planted our boots on the seat of his pants and shoved him forward on top of the first pilot, following in hot pursuit.

In a great confusion of arms and legs we managed to overpower them both, and handed them over to Brown and Dunsmore, who were now armed with wrenches, at the back. In the scuffle we had lost sight of the Beretta pistol, so I told Wilkie to look for it while I turned my attention to the second pilot, who was just about to land the aircraft on the sea. I immediately shoved the revolver under his chin to indicate to him to lift her up. He did what he was told, and promptly climbed to about 1,000 feet. In the meantime, Wilkie had found the Beretta automatic so I instructed him to stand guard next to me facing aft, and I slid into the first pilot's seat and took the controls.

I shoved the nose down to fly below radar detection at 200 feet, the concerned Italians indicating in gestures, and exclaiming, '*Non troppo basso, non troppo basso, l'idrovolante*!' (Not too low, not too low, the floats). Then, thinking we were heading for Africa, they shouted nervously, '*No, Africa, no benzina*!', to which I replied, 'Not Africa, Malta!'

Panic-stricken, they cried out, 'Spitfires, Spitfires!'

'Ja, well,' I am thinking, 'first, we have to find Malta.'

Once level and cruising, I looked down at the instrument panel in front of me and found an unfamiliar field of illegible words and decimal points. While certain flying instruments are self-explanatory, all the switches and gauges

WHERE THERE'S A WILL, THERE'S A WAY!

were labelled in Italian and in metric measure, not imperial. Aware that we had been destined to go on a short flight to Taranto, we realized that we would at some stage need to change the fuel tanks. This would require help from one of our captives, the one most likely to be easily intimidated. I reckoned this would be the flight engineer. So I handed the controls over to the second pilot and ordered that the engineer be released and brought up front. Then, in pantomime and with a waving revolver, I made it clear to him that if we ran out of gasoline I would shoot him. If there were any fuel tanks to be changed he had better do it without any nonsense. He said, '*Capisco, si, capicso*' (I understand), and was sent back to Brown's guardianship.

I then called to Dunsmore, 'Bill, look for maps or charts; anything that will help us navigate our way back to Malta!' And then I again took over the controls from the second pilot and sent him to the back, to Brownie. It wasn't long before I spotted a dot in the sky approaching from port, and realized we were on an intercept course with another aeroplane. What then happened I suppose one could call it inverted luck. As soon as I recognized it as a German Ju52, I whipped the red tabs off my epaulets so as not to be identified.[1] I waggled my wings in friendly salute, which he reciprocated as he passed about 200 yards behind us.

After a thorough search Dunsmore reported, 'No maps or charts here, Ted.'[2] So here we were, flying in a strange aircraft, with no maps or charts, an illegible control panel and not a clue from where we had set out or which way to head. I was left with some serious guesswork to do. I decided to turn due west, hoping to see the coastline of Italy. I reasoned that I could edge down and around Italy towards Sicily; from the south-east point of Sicily I would fly the course of 220 degrees to Malta. This course, in the event of wireless or navigation failure, had been drummed into us at briefing, and I was pleased for it now. We flew for what seemed a very long time. Then I spotted some mountains in the distance, which I hoped were on the tip of Italy. I immediately dropped right down to about 20 to 30 feet above the water, again as a precaution against enemy radar. I altered course to south-west, until I picked up what I supposed was Sicily. A heavy haze restricted my visibility, forcing me to fly about 5 miles from the coast.

At about twelve o'clock the engineer officer started making signs, was released, and was brought up to the front. This was a harrowing moment. We were now vulnerably close to enemy territory. If he were to take a chance and shut everything down, we would have been bloody well buggered! I watched, gun in hand, my nervous apprehension hidden under a stern gaze, as he turned various fuel cocks. We flew steadily on and he was duly discharged. With my visibility now at about only 4 miles, I took a chance, and at the first point of a large bay I came across, I set course of 220 degrees. A few minutes later, the coast reappeared. Yet another sweeping bay confronted me, so I set course again. Soon there was yet another sweep of bay in front of me. By now I was sweating and saying to myself, 'Where the bloody hell am I? I might still be up in Italy.' However, on the fourth attempt to set course we flew on and on, and no more land appeared. This was a relief, in spite of the fact that, having guessed my way across half the Mediterranean, I was by no means sure that it was from Sicily that I had set course. 'Keep going, just keep going,' I repeated to myself, hoping the fuel would hold out.

Two problems now loomed large – finding and recognizing Malta and, if we did, the bloody good chance, of being set upon by Spitfires. I shouted, 'Keep a damned good lookout for the Spits,' to my crew, and decided that, in the event of sighting Spitfires, I would immediately get the Italian first pilot back at the controls and order him to alight on the sea. Well, this was the plan in theory. ... We flew on hopefully, and eventually an island appeared on the horizon. As we were never allowed to approach Malta from the north, I had never seen that side of the island, which emerged before us with none of the familiar landmarks. In a quandary, I'm questioning to myself, 'Is it Malta? It's got to be Malta.' I saw another small island on the starboard side, and hoped it was Gozo.

Then there was a yell from Dunsmore: 'Look out, Spits!'

Indeed, it was Malta, and there were four Spitfires in line astern at 1,000 feet on our tail, with the first one dipping his wing to swoop on us. I grabbed the first pilot, shoved him in his seat, and indicated to him to land her. The second pilot had already started to throttle back, in anticipation of attack, when the first Spit opened fire directly in front of us at about 30 yards. The second chap got his aim down to 15 yards in front of us, as we were slowing down rapidly now. To get into wind, the first pilot made a gentle turn to starboard and, as he did, a rapid run of bullets ripped through the port wing. The third Spit had caught us.[3] Dunsmore, in the meantime, had whipped off his white vest and was waving it out of the top hatch in surrender. As we touched down, the fourth Spit held off, and I commended the Italian first pilot for his calm and careful handling of the alighting under attack.

On Malta, Herschell Reilley, a Canadian navigator, was convalescing in the hospital at St Paul's Bay. He was standing with a few other patients on the hospital's long balcony, which overlooked the famous and sacred bay in which St Paul was shipwrecked in AD 60. This is what he wrote about the event:

> As we watched the sky, as was our habit, one of the group pointed and said, 'Hey, what's that coming in there?' We all looked and knew, from our aircraft recognition course, that it

WHERE THERE'S A WILL, THERE'S A WAY!

The incredible X-ray photograph of the Cant landing on St Paul's Bay while under attack. This photograph was taken by someone who, while watching the drama unfold from a hospital balcony, had had the forethought to grab some X-ray film to take the picture. The image is of the last Spitfire honing in on the Cant as it alighted on the sea. Author's collection

was an Italian Cant Z.506B. About five seconds later someone remarked, 'And he's not alone!' there were four Spitfires above him. We all agreed, as someone said, 'They won't shoot it down, it's only a transport plane and very lightly armed.' As they spoke, the first Spit peeled off, came close and low and opened up on it, and drew a spray line with bullets directly in front of it in the water. On the third assault the Cant landed. 'So who is it?' the question was on everyone's lips. Someone joked, 'It's Mussolini come to surrender!'

Meanwhile, having alighted safely and now stationary on the rough sea, we all climbed out onto the wing, and in due course an air-sea rescue launch appeared. The skipper emerged and I suggested we start the plane and taxi in. When I indicated to the pilot to start the engines, he looked at me and shrugged. *'Non possibile, no benzina.'* This, amazingly, proved to be a fact – the tanks were bone dry. The unknown amount of fuel on board had brought us to a mile off Malta into St Paul's Bay, and we had incredibly been brought to safety on the last drop of petrol!

Well, no power meant being towed in by the rescue launch, but the plane

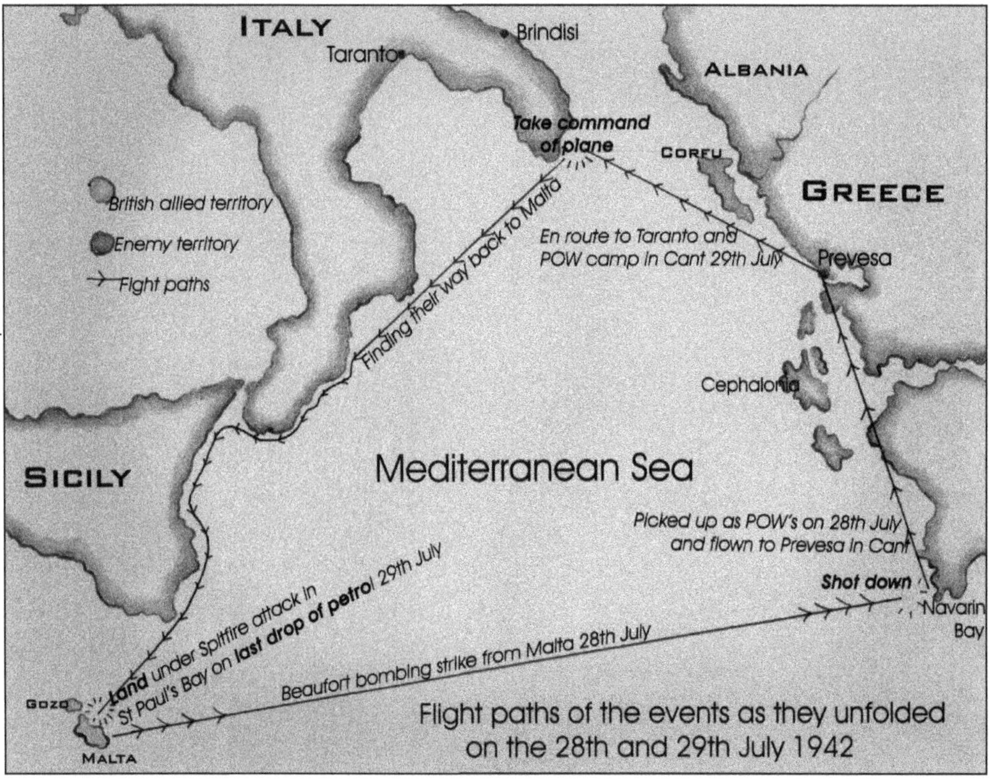

Map of attack, ditching, Prevesa and hijacking.

The Cant with tug and crews being towed into St Paul's Bay, Malta, 29 July 1942.
Author's collection

WHERE THERE'S A WILL, THERE'S A WAY!

Arriving at the quayside of St Paul's Bay, Malta, Wilkie standing in the foreground, 29 July 1942. Courtesy of the Roy C. Nesbit collection

Mooring at the quayside, Malta, 29 July 1942. Courtesy of the Roy C. Nesbit collection

stubbornly resisted the pull by persistently wind-cocking and rocking wildly on the rough cross-current on which we had landed. We did eventually get some forward momentum and slowly approached the quayside, passing the sacred site of St Paul's Island on our way. Awaiting us was a large crowd of jubilant cheering Maltese giving us a heroes' welcome. When our Italian prisoners disembarked onto the quayside I was alarmed when the angry Maltese crowd rushed towards them. I had to very smartly step in, order them back and brandish the Italian Beretta pistol to protect them from the mob.

In his correspondence to me, Steve Haffenden, the coxswain of the sea and rescue launch that brought Dad, his crew and their Italian prisoners in from the bay to the quayside, remembers:

> It was a lovely sunny afternoon when the rescue launch 107 was called out to investigate an aircraft on the water. As the launch approached the floatplane I noticed as well as the crew, somebody was standing on the wing waving something white.

Ted and his crew with their Italian prisoners, St Paul's Bay, Malta, 29 July 1942.
Courtesy of the Roy C. Nesbit collection

WHERE THERE'S A WILL, THERE'S A WAY!

We took that as 'surrender'. After we had tied up alongside the jetty at St Paul's Bay, and put all those from the Cant flying boat ashore, Strever asked me where the officers' mess was so that he could give the Italian Cant crew a drink as the Italians had treated them very well while at Prevesa. I had to disappoint Strever, as there was no officers' mess at our base. The Italian armed escort's jaw was slightly disfigured from the blow he had received from Wilkinson, and I remember Strever telling me he felt sorry that they had had to hurt the Italian. Army transport was awaiting our arrival back at base and they were all transported, to where, I do not know. HQ, I guess! The Beaufort crew came ashore feeling in a state of excitement, and those happy smiles I'll never forget.

'Later, I tried to reciprocate the Italians' hospitality with a meal before they were taken off to prisoner of war camp. While they appreciated the gesture, I know they weren't impressed with the bully beef fare of Malta,' said Dad.

On 30 July 1942, Squadron Leader Gibbs, who had been off duty on the previous day, wrote:

I knew Malta well enough to expect something of moment to have occurred in our absence; ... but the sight that met my eyes was beyond the mundane realms of my conjecture: four grinning ghosts were haunting the office, the shades of Strever and his crew. A second of speechless doubt passed, then the ghosts came to life and I found myself listening to a story such as I never thought could exist outside fiction. ... For many days afterwards the island buzzed with the story, and the squadron's joy knew no bounds; that night we entertained the returned captives in the mess, the next day I sent them on a week's leave. This spectacular event was a fitting ending to a week of successful attacks. ... Optimism coloured the letter that I wrote that evening; it said that the first stepping-stones had been firmly laid and that the path would be complete before Christmas.

Wing Commander Patrick Gibbs, *Torpedo Leader*

For me, the auspicious landing of the Cant in St Paul's Bay, which was named after St Paul's shipwreck there in AD 60, has an intriguing and

The Italian POWs being led away by British personnel, Malta, 1942. Courtesy of the Roy C. Nesbit collection

The Italian prisoners arriving at Waterloo Station, London, 1942. Courtesy of the Roy C. Nesbit collection

WHERE THERE'S A WILL, THERE'S A WAY!

beautiful correlation to the biblical account (Acts: 27: 39) of the shipwreck. Dad's account states, 'I couldn't recognize Malta,' and the biblical account states, 'we knew not the land'; St Paul's sailors knew Malta from previous journeys but both were approaching Malta from the unfamiliar side. The strong cross-current of St Paul's Bay that Dad talks of is, in fact, the place where St Paul's ship was wrecked, as written in Acts, 'where two seas met.' On the small limestone island in the bay, a solitary statue of St Paul stands on a huge plinth, facing the sea with his right arm outstretched and raised – the gesture of divine blessing.

A few weeks after the skyjacking, Harry Coldbeck, a pilot and reconnaissance photographer at Luqa airbase, was asked by Air Headquarters in Malta to deliver a package addressed to Regia Aeronautica. This was to be done by dropping it out of his plane at Catania airbase in Sicily, a dangerous proposition since it was an enemy airbase. The package contained all the personal effects of the Italian crew captured by Dad and his crew, being returned to the Italians' next of kin. Harry got his men to make the parcel visible by tying long brightly coloured fabric streamers to it.

He then went on to carry out a photographic mission, and on his way back to Malta he took a detour towards Catania. Once he was sure there was no defensive action from the airbase he descended, slowed down, opened the hood of his Spitfire and hurled the parcel out of the plane. A few months later, he himself was captured by the Italians, and discovered after asking that the parcel had been found and the belongings returned to the families of the Italian crew members.

Chapter 15

Doing the Flying Cha-Cha-Cha

*Then pray that the road is long
That the summer mornings are many
That you will enter ports seen for the first time
with such pleasure, with such joy ...*
C.P. Cavafy, from the poem *Ithaca*

'Sir, I want to be relieved of my duty; at thirty-two, I'm too old for this kind of work. I have done three trips with this crew, have been shot full of holes on two trips, and shot down on the third!' Dunsmore appealed to his commanding officer, directly after the skyjacking escape.

Dad continues the story: 'The CO agreed and relieved him of his duty, and sent him back to the UK. So I was now without a navigator, until a Canadian navigator by the name of Herschell Reilley, who had been recuperating in hospital, joined our crew and we were ready to move on.'

Herschell relates his side of events on leaving hospital:

> The transport came and we set off for the airbase at Luqa, but went in the wrong direction. We turned in towards some stone buildings (what else in Malta?) and stopped in front of a massive entry. On our querying where we were, the driver said, 'It's the officers' rest camp.' I went to my room shortly after dinner with a view to spending some luxury time in a proper bed instead of a camp cot. It didn't work out though, as on the way I met a lad called Ted Strever. Upon enquiring why he was there he related what had happened a few days earlier. He was quite surprised when I related my view, from the St Paul's

DOING THE FLYING CHA-CHA-CHA

hospital balcony, of the last act of the drama. The next morning we both returned to our airbase, Luqa, together. We found that the remains of the squadron had moved to Egypt, so we were there on our own. On our arrival at Luqa we were met by the CO, who acknowledged us and said, 'Oh, I'm pleased to see you two have met. As Dunsmore has returned to the UK, Herschell, you will take his place as navigator on Ted's crew.'

Herschell Reilley, personal correspondence in 1997 to the author from Canada

Herschell Reilley was a little apprehensive at this news, knowing Dad's rather daredevil reputation. He wrote in a letter to me:

Well, I knew what he was like! And decided when going into a strike I would just turn my back, not look and get on with my job.

Dad again ...

In August, 217 Squadron was posted to Egypt, to an airbase called LG224 in the desert, situated very near the three pyramids of Giza near Cairo. We were the last crew to leave Malta. I was still a lieutenant and had by now received the news that I had been awarded the DFC for the Cant episode. The day before we left, we rambled about and around piles of rubble at Luqa. Most of the streets were blocked by the devastation left from the bombing. Later, we went to the docks to see the famous oil tanker, the *Ohio*, which en route to Malta had been attacked day and night for about two and a half days by enemy submarines, dive-bombers and medium bombers. It caught fire at one stage and the crew abandoned ship. However, because the fire remained on the upper deck and didn't reach the fuel, she was re-boarded, the fire was put out and, listing badly, they attempted to keep going under this unrelenting assault from all sides. There was a stream of Spitfires going out to try and save this tanker because the fuel she was carrying was absolutely vital for Malta. I can't think of anything more courageous than those sailors on that ship. Eventually, aided by a destroyer and a minesweeper, she was towed into Malta. She entered Grand Harbour, sandwiched between the two vessels, with cables under her, which were slung between them, to keep her from sinking. The cheering and applause from the crowds on the quay, as they came into the harbour, was a very moving and exciting experience. Her arrival saved Malta, at a crucial time, from the fuel crisis and thus from grounding and shutting down all of its defences.

ON LAUGHTER-SILVERED WINGS

The next day, we, the last crew of our squadron to leave Malta, made ready to leave for the last time. Our aircraft was fitted with auxiliary fuel tanks, and all our personal kit was loaded. We took off from Luqa, loaded to the hilt. As usual, once airborne I released my safety harness as it restricted any comfortable movement. Within seconds, the bloody starboard engine shuddered, coughed, burst into flames and seized. I had no height and no time. I hit the fire extinguisher button and started to turn around. Malta consists of hundreds of small fields, all of which are surrounded by 3-foot to 4-foot limestone walls, so not a chance of anywhere to land. Herschell, who saw me struggling at the controls to get the plane back into the circuit, leapt up from his station, grabbed my harness straps and got me hooked up. The nearest airstrip was at Ta' Qali, the fighter airbase, and I was aiming for it. I just managed to do the circuit, with wheels and flaps down, and we swept in, missing a gun emplacement by about 6 inches, and hit the ground heavily; there was no bounce this time. Herschell, who, having secured me in my seat, didn't have time to get back to his, had braced himself next to me and hung on to a handle next to my seat. On impact he was thrown forward, wrenching his right shoulder joint. I fought the controls on landing to keep her on her wheels and we eventually stopped.

In a letter to me in 1997, Herschell writes:

> I looked at Ted. He was still sitting fastened in his seat, looking around to see how he had done. He suddenly flung off his restraining straps and bolted for outside, yelling at us to get out as the plane might blow at any moment. We did – it didn't.

Dad continues ...

Thank God, the inbuilt fire extinguisher had doused the flames before we came down. We had about a hundred gallons of fuel on board, and if that had gone up we would have had it! Apart from Herschell's shoulder injury, no one was hurt, but the aircraft was quite a mess. The engine was burnt out, the tail wheel was bent and broken, and the main wheel had burst and was shredded. The hard landing had buckled the wings and because of the weight of the fuel, the main frame had bent in the middle, just behind the turret. And here we were, back in Malta!

We loaded our kit onto a truck and returned to Luqa, to the mess. There we spent the afternoon talking to other crews who had arrived. The next day, we boarded a Dakota, this time as passengers, to fly to Egypt. We took off, only to find out that two of the gauges on the control panel weren't working. In an

DOING THE FLYING CHA-CHA-CHA

Ted and the wreck after crash-landing at Ta' Qali, Malta, 1942. Author's collection

attempt to get them to read correctly, the pilot knocked on them with his knuckles a few times, swore loudly, and turned the aircraft around to return to Luqa. It was now becoming a bit of a joke at the mess that we just didn't want to leave the place.

On our third attempt to leave Malta we finally made it. We were put onto a Dakota, which was loaded up with kit bags. We threw ours in and climbed on top of the heap. This meant we were sitting high above the window level, so had no outside view. With nothing but the familiar sight of each other to look at, our attention fell upon the pile of kit bags and luggage we were sitting on top of, each with a tag on it. It didn't take long for us to each have a tag in our hand reading the address. All were civilian addresses. Slowly, a shocking realization hit home: we were sitting on the personal effects of those who had not made it. It was a sobering and disturbing thought. At that moment it wasn't only the engine noise that made it impossible to talk, as each of us sat, speechlessly submerged in our own thoughts. It was a meditative trip shadowed with ghosts.

We set down at LG224, a barren desolate place in the middle of nowhere, literally just a landing strip cut out of the desert.

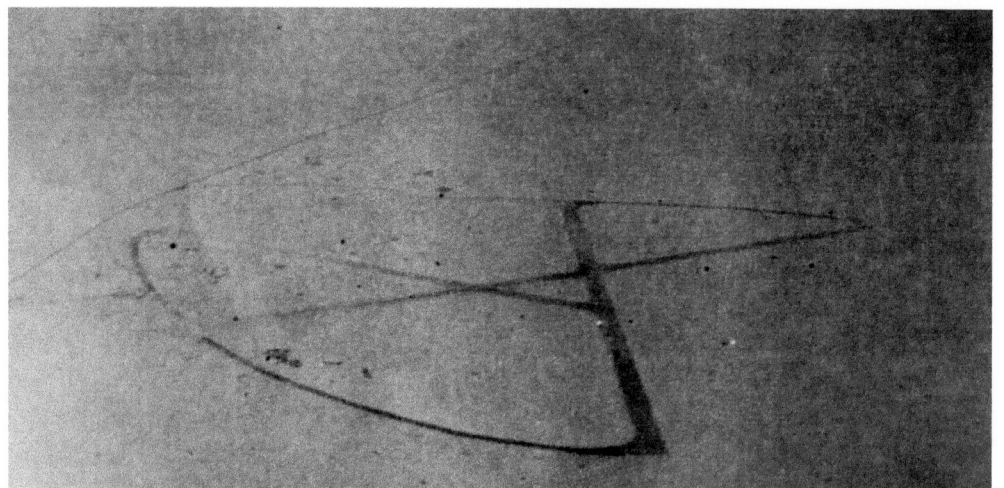

LG224 runways just south-east of Cairo with the pyramids of Giza a couple of miles north-east in the desert. Author's collection by courtesy of Herschell Reilley

It was really used as a refuelling stop and a sort of staging post, so, until posted elsewhere, there was nothing to do but hang around. Luckily we only stayed there for one night, in a dormitory. The next day we discovered Cairo, and moved into the New Zealand Club that afternoon. The first thing I did was to go to Air Force headquarters at Villa Victoria, with a special request to see my dad. I knew he was stationed in the desert somewhere near Cairo, with No. 7 Wing. He was the safety equipments officer, or NCO, and had the vitally

Taking off from LG224 with the pyramids of Giza to the right of the plane. Photograph taken from Ted's aircraft. Author's collection by courtesy of Herschell Reilley

DOING THE FLYING CHA-CHA-CHA

responsible job of making sure that parachutes were not damaged and packed perfectly, and that all items of essential equipment were in perfect working order. They were intrigued to know that my father was serving there and so, to surprise him, they sent a message to the CO of No. 7 Wing, merely saying, 'Air Sergeant W.H. Strever is to report to headquarters at 9.00 am tomorrow morning.' Well, poor Dad spent a sleepless night, because being summoned to HQ usually meant trouble, and he wondered what trouble he was in!

The following morning at nine I waited for him in an office especially vacated for our surprise meeting. I'll never forget the expression of amazement on his face as he walked in and saw me standing there. We hadn't seen each other for two years, and we just stood there for a moment hugging each other. To my surprise, the CO had arranged for us both to have a few days' leave so that we could spend some time together. It was wonderful. We went sightseeing around Cairo and of course saw the pyramids and the Sphinx, and had a really grand time. I also got permission to take him for a flip in a Beaufort over Cairo, which delighted him immensely. Then I went back to his squadron with him and met some of the boys there.

Ted flying his father, Pop, over Cairo in a Beaufort, 1942. Author's collection by courtesy of Herschell Reilley

ON LAUGHTER-SILVERED WINGS

Oh my, there was Dad in the middle of the desert, yet again making a plan! While everyone else was sleeping on the sand, he – being the old soldier that he was – had somehow set himself up with an iron bedstead. It had a proper mattress, replete with sheets, and it stood on the sand in his tent. He proudly insisted that I sleep on it, in proper sheets, while I was there! In fact, he had also actually built a caravan there in the desert, with some clever resourceful inventions, for the colonel of the wing, Dougie Loftus. Well, that was my old man, the greatest improviser!

Our leave ended. Dad had to get on with his work, and I had to report back to Cairo. The crew and I would end up staying for another two weeks, waiting for our orders to leave for Ceylon. This was the nature of the war; our lives were spent either in the middle of violent activity, or sitting around waiting for something to happen. So, with not much else to do, we familiarized ourselves with life in Cairo.

Dad would never have divulged to me, his daughter, the details of 'life in Cairo', but I was enlightened through my later (1997) correspondence with Herschell Reilley.

According to Herschell, he and Dad, Wilkie and Brownie were having a drink in their favourite bar one night, just passing the time, when an Arab approached and introduced himself.

'*Sahibs*, you want to see the Cairo night life?'

They looked at each other and Dad said, 'You know, chaps, we haven't seen much of this famous Cairo nightlife. Maybe we should give it a go.'

They turned to the Arab. 'What is it all about?'

He started his sales pitch. 'You like to see young ladies, not so young ladies, when make love? Well, you come and I show you. Not far, you see private show, no one see you.'

Dad turned to the barman and asked, 'You ever seen this production he's talking about?'

'No, Sir, but I hear it is very entertaining. You might do well to view it!'

Apparently, Dad turned to the chaps and said, 'In other words, he doesn't know!'

Brownie then, quite eagerly, said, 'We've nothing to lose,' and turning to the Arab asked, 'How much?'

'Oh, very small amount,' the Arab assured him.

Brownie haggled with him for a while and they finally agreed on a price. The businessman said, 'You follow me, gentlemen.'

With that, they all walked out of the door and followed him through the

DOING THE FLYING CHA-CHA-CHA

Ted (to Dad with best love from Ted) in Egypt at the tented camp in the North African desert, 1942. Author's collection

South African newspaper cutting of Ted and Pop's surprise meeting in Cairo, 1942. Author's collection

S.A.A.F. Father and Son Meet in Desert

S.A. PRESS ASSOCIATION'S WAR CORRESPONDENT

Cairo, Monday.

A FATHER and his son, both in the uniform of the South African Air Force, have just met at S.A.A.F. Headquarters in the Middle East after being separated for two years. Both joined the S.A.A.F. on January 15, 1940.

The father is Air-Sergeant W. H. Strever, who has a fine record of active service in both wars, and his son is Lieutenant Edward Theodore Strever, who has just been awarded the Distinguished Flying Cross. He is a well known South African sportsman. Special leave was granted to them so that the meeting could take place.

Before the war, Air-Sergeant Strever was the proprietor of a motor garage in Klerksdorp and his son was in a chartered accountant's office.

In the Great War the father served with the Pretoria Regiment in South-West Africa and with the artillery in East Africa, Egypt, Palestine, France and Belgium. He enlisted again in this war and went to East Africa in July, 1940, and went through Italian Somaliland and Abyssinia with his squadron.

After spending some leave in the Union, Air-Sergeant Strever came to the Middle East in April this year with a fighter squadron. He found time to build a caravan in the desert for his commanding officer at a cost of 18s.

Despite his age, this veteran soldier is one of the fittest men in the squadron.

LIEUT. EDWARD THEODORE STREVER

Lieutenant Strever is a torpedo-bomber pilot, who has attacked Italian warships and convoys in the Mediterranean.

In operations on June 15 against a strong force of Italian battleships, heavily escorted by destroyers, Lieut. Strever closed, in spite of anti-aircraft fire, to drop a torpedo at one of the enemy battleships. On another occasion in the central Mediterranean the crew scored a direct hit with a torpedo on an enemy merchant vessel proceeding to North Africa.

His mother has left Pretoria to stay in Johannesburg with her young daughter.

Pop and Ted at the entrance of the Mortuary Temple at Giza, Cairo, 1942. Author's collection

Pop Strever with the caravan he made in the desert for his CO, North Africa, 1942. Author's collection

Ted and Pop at the tented camp in North Africa, 1942. Author's collection

dark streets of Cairo until, after about five minutes, they arrived at another door. He opened it, ushered them in, and took them to an upstairs room, which was large and empty but for a row of chairs on one side, all facing a wall. In front of each chair, at eye level, was a small shutter that slid open and shut over a hole in the wall. They sat down, opened the little shutters and peeped through the holes. They revealed a room decorated with murals and drapery around a large bed, behind which was a large mirror with a door on each side of it. As they watched in anticipation, one of the doors opened and in walked a young woman, covered from top to toe in a long purple robe. Behind her she pulled a 6-foot seaman into the room. After the two had disrobed with some ritual, she beckoned to him invitingly with her forefinger, which, for some time, resulted in a game of cat and mouse around the bed (this was designed to heighten his appetite and extend the show). After this frolicking, the object of his lust eventually lay waiting on the bed and with an urgent whoop, he was attending to business. When the coupling was complete, she slipped off the bed and ran out of the room, waving at the peepholes as she disappeared, to the click of a lock. The seaman recovered himself, got dressed and also left the room.

They left the theatre and returned to the bar to report back, which they did in some detail, for the price of a drink. The next two days they were flying on exercises, so only returned to the bar three days later. While they were sitting there, in walked Charlie Evans, who they knew from 42 Squadron, and he joined them.

'What have you been up to lately Charlie?' they asked him.

He looked around the group, leaned into it and confided, 'Have you ever run into a free night, when everything was provided. I mean … everything?'

'No,' they replied.

'Well, I did last night! I was at the Shallufa Bar, a couple of streets up the road, and was invited to have a drink with a local girl who bought me a whisky. As we sat chatting she said, "How you like to have party with me, just you me?"' At this, Charlie said, 'I gave her the once over, and decided to take up her offer. We walked a few blocks, and entered a building, went up to an upstairs room, with a bed …'

They all knew what was coming next! By now, they were all grinning, some of them turning away to stave off bursting into a fit of raucous laughter. Charlie, looked at their knowing grins and said, 'What's up?'

Brownie chipped in. 'Shall we tell him? The poor bugger should be told.'

The others nodded, so Brownie continued to Charlie, 'Think about it, man! You don't get nothing for nothing, especially in this country. Too bad you weren't three days sooner or we were three days later.'

'What are you talking about?' asked a confused Charlie.

'Then we could have seen a performance starring you, old son!' replied Brownie.

'What?' exclaimed the alarmed Charlie.

'Didn't you wonder why the room was so large? The waist-high, carved wooden decoration around the walls has a row of holes hidden in it, and on the other side of the wall, there is a bunch of servicemen peeping at the proceedings through the holes!'

There was sheer panic on Charlie's face at the thought of this horrible truth. He left the bar, muttering to himself, and they never saw poor Charlie again. They all agreed that, should someone say to him, 'Didn't I see you in that girlie show?' he should lie through his teeth and say, 'Just got in yesterday.' Two days later, they left Cairo.

They were then sent to Wadi Arah in Palestine, a repair depot, to deliver their plane and to collect a Beaufort M11[1] to be flown to Ceylon. None of them wanted the posting to the Far East, and tried their damnedest to get out of it but with no luck. On 10 October they flew from LG 224/Cairo across Israel, the Dead Sea, along the main pipeline to Baghdad, and landed at Habbaniya in Iraq. Habbaniya was an RAF station in the desert, some 60 miles west of Baghdad. They could see Baghdad, across the Tigris River, in the distance.

About Habbaniya, Herschell Reilley writes:

> It was here that I discovered that great areas of the desert is dry earth, and not sand. Being a bunch of Brits ... they had developed the place into an English village, complete with hedges and flowerbeds. Lawns were dark green. This was accomplished by making a boundary wall about 8 inches high in cement around the beds and lawn. Wells were drilled and pumps distributed the water. Lawns were flooded to about 6 inches deep every three weeks. They even had English trees. It was a remarkable sight. The paved streets completed the picture. We slept in the transit house; each of us was given our own room, with a complete batman service. Breakfast was quite affluent, which made one wonder where the war was.

DOING THE FLYING CHA-CHA-CHA

Dad continues the story …

The next morning, we reluctantly left this little English Eden, and flew east, across the desert, over the Persian Gulf to the island of Bahrein. Bahrein consists of two main islands, the airstrip being located on the smaller of the two, and was extremely flat and smooth. In the briefing before leaving Habbaniya, I was warned: 'Ted, when you take off from there make sure you are airborne and at a height of least 100 feet before you retract the undercarriage.' Well, I didn't believe this until we landed there. Somehow there was no sensation that the wheels were running on the ground. The sand runway had water beneath the surface, which kept the sand damp and firm. This produced a beautiful velvety cushioning effect, even when walking on it. However, it was dangerously deceptive when taking off. There were three aircraft lying at the end of the runway with their wheels up, to dispel any doubts about this.

Herschell writes:

> We spent the afternoon at the market, which in this wealthy emirate spanned an area of over 20 acres, the most extensive market we saw anywhere in the East. It offered everything from camel hair luggage to strings of pearls, and we marvelled at the selection of goods. Unfortunately, we hadn't been paid since leaving Malta, so were forced to just look. However, it was fun to feast our eyes on all the merchandise.

Dad and his crew seemed to truly be taking their own journey to Ithaca, which was immortalized by the Greek poet Cavafy in his beautiful poem of the same name:

A Spitfire taking off from the sand runway at Bahrein. Author's collection

ON LAUGHTER-SILVERED WINGS

With such pleasure and such joy
Stop at Phoenician markets
And purchase fine merchandise,
Mother of pearl and corals, amber and ebony
And pleasurable perfumes ...

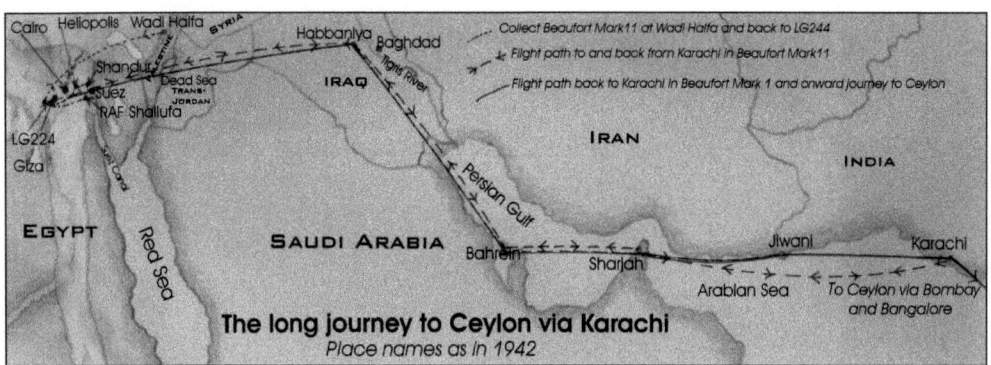

The Middle East – the long journey to Ceylon via Karachi.

The next morning we were flying over the Persian Gulf again on our way to Sharjah, which is on the eastern coastline of the Gulf. We flew past the town, and a few miles inland we landed at the RAF base at the Sharjah Fort [now Al-Hisn Fort]. It was such a surprise to see this pukka fort, which rose up from the desert. It had corner towers, joined together by huge walls topped with ramparts. The rooms and halls ran around the inside of the wall surrounding a large inner courtyard. It reminded me of the typical old Foreign Legion style forts of the Sahara Desert. We entered through the enormous doors at the entrance and they were locked behind us. The plane was guarded by a local Arab who was barefoot and wore a large white turban on his head, and a kukri (a curved dagger) at his belt. He looked wonderfully exotic and proud. At night he slept on the sand between the wheels of the plane, armed with a rifle.

He reminded me of a night in Egypt, at LG224, where our planes had been similarly guarded. Hersch and I had been standing chatting outside our digs, enjoying the stillness and desert air before retiring for the night. There was pitch-black darkness over the desert, and in the silence, suddenly we heard a shot, then, a few seconds later, a shout of, 'HALT!', followed by utter silence again. We looked at each other, and I said to Herschell, 'Shouldn't that be the other way around?' He agreed. The next morning they had picked up a body and we all agreed that going to fetch something from the plane after dark was not a good idea!

DOING THE FLYING CHA-CHA-CHA

Sharjah Fort from the air on take-off. Photograph taken from Ted's aircraft. Author's collection

Sharjah Fort's main gate leading onto the runway, 1942. Author's collection

Arab plane guard at Sharjah, 1942. Author's collection

DOING THE FLYING CHA-CHA-CHA

The next morning we flew the five-hour trip from Sharjah Fort to Karachi, which is on the northern shores of the Arabian Sea, just west of the Indus River delta, in that part of India that is today part of Pakistan. As I alighted, an engineer officer approached me, and pointing at the plane, he said, 'What is this?'

'What do you mean, what is this?' I replied, thinking, what an idiotic question.

'What type of aeroplane is it?'

'Oh, it's a Beaufort,' I answered, puzzled.

'What kind of engines are those?' he asked.

'Beaufort Mark 11, Pratt and Whitney engines,' I said.

'Well,' he said, 'you can get back into it, and take it back to the Middle East. We don't have Beaufort Mark 11s here, and we've got no Pratt and Whitney engine spares!'

Somebody at headquarters had slipped up. We shrugged and, knowing we would have to wait for an order to fly back, we went into the airport for debriefing and some lunch.

The airport offered the finest of services. It was efficiently run, lavishly decorated and offered the best of everything. Lunch there was splendid, in all senses of the word. Of this experience, Herschell writes:

> Karachi was a very interesting city. We landed at a first-class airport. Someone was there to point us to the parking spot, and a vehicle was waiting. A crew unloaded our travelling gear, and a guard was put on the aircraft. We then headed towards a very modern terminal, which centred round a tower some 150 feet across. As we crossed the floor spanning the tower, we stopped and looked up. The architecture was superb, with decoration all the way to the top, the whole area covered with an intricate glass dome. We were conducted to an office and debriefed. Lunchtime came, we entered the restaurant. The walls were covered with Indian murals, there were antiques throughout, and oriental rugs covered the floor. The large teak tables were surrounded by armchairs, which had woven (rattan) seats topped with padded cushions. Crystal chandeliers provided the overhead illumination, while elephant-head lamps with their trunks up supported shaded illumination all along the walls. The table attendants had their hair up in a bun. They were dressed in white linen jackets, buttoned to the top, a floor-length sarong and

walked barefooted. It was 'first class'. This was the first time I had seen the large white linen napkins displayed in the form of various birds and animals. I marvelled at the skill that could produce this art. I finally spotted the person who was responsible for producing them. He seemed to be well past sixty and was dressed like the others. He kept a keen eye on the tables; when new napkins were required he would approach the table with clean napkins and, with very nimble fingers, would fashion them into a zoo of birds and animals. I watched with great fascination at how very skilled he was at his craft.

After enjoying the peaceful ambience of the restaurant and the most satisfying meal, we left the airport. Now, the Middle East had been very different but fascinating for us, but what confronted us on the streets of Karachi was a jarring rude assault on every one of the senses. It was a shock to my English, middle-class, hygienic sensibility. Potent unsavoury smells filled the air and had me gagging. The filth that covered the pavements was too ghastly! They were splattered with red betel juice spittle and the locals would squat on the pavement to defecate. We hadn't been keen to come to the East and, my God, this place put the seal on it!

Another of the disturbing sights, albeit an unusual necessity in Karachi, was the en masse public removal of dead bodies from the streets each morning. This was done by a camel wagon, which pulled a large box cart made of planks. Every morning at six o'clock these wagons would pass through the streets, picking up the bodies of the homeless souls who had died in the night. There wasn't much dignity to this process and the bodies were hurled onto the cart like mielie sacks. The full load was then taken to the banks of the Lyari River, where a pyre would be built in alternating layers of cordwood and bodies. The final size of the pyre was usually about twelve layers. This would be set alight and when it had burned down to a pile of ashes the whole lot was scraped off the shore into the river. This, we learned, was a practice of the Hindu religion, a religion that focuses on the spirit of man, rather than the flesh. These practices got me thinking!

After a few days, we climbed back into our aeroplane and headed back towards Egypt. So we didn't reach Ceylon this time either. To alleviate boredom and add to our list of places we had visited, we decided to do a bit of a detour, so en route we landed in more places than were really necessary, just to see them. On arriving at LG224 we were shunted off and told to go to Heliopolis (Cairo), where they told us to go to Shandur, on the Suez Canal. Finally, at Shandur, they didn't shove us off anywhere else.

And we soon realized that the officers there thought that we had come

DOING THE FLYING CHA-CHA-CHA

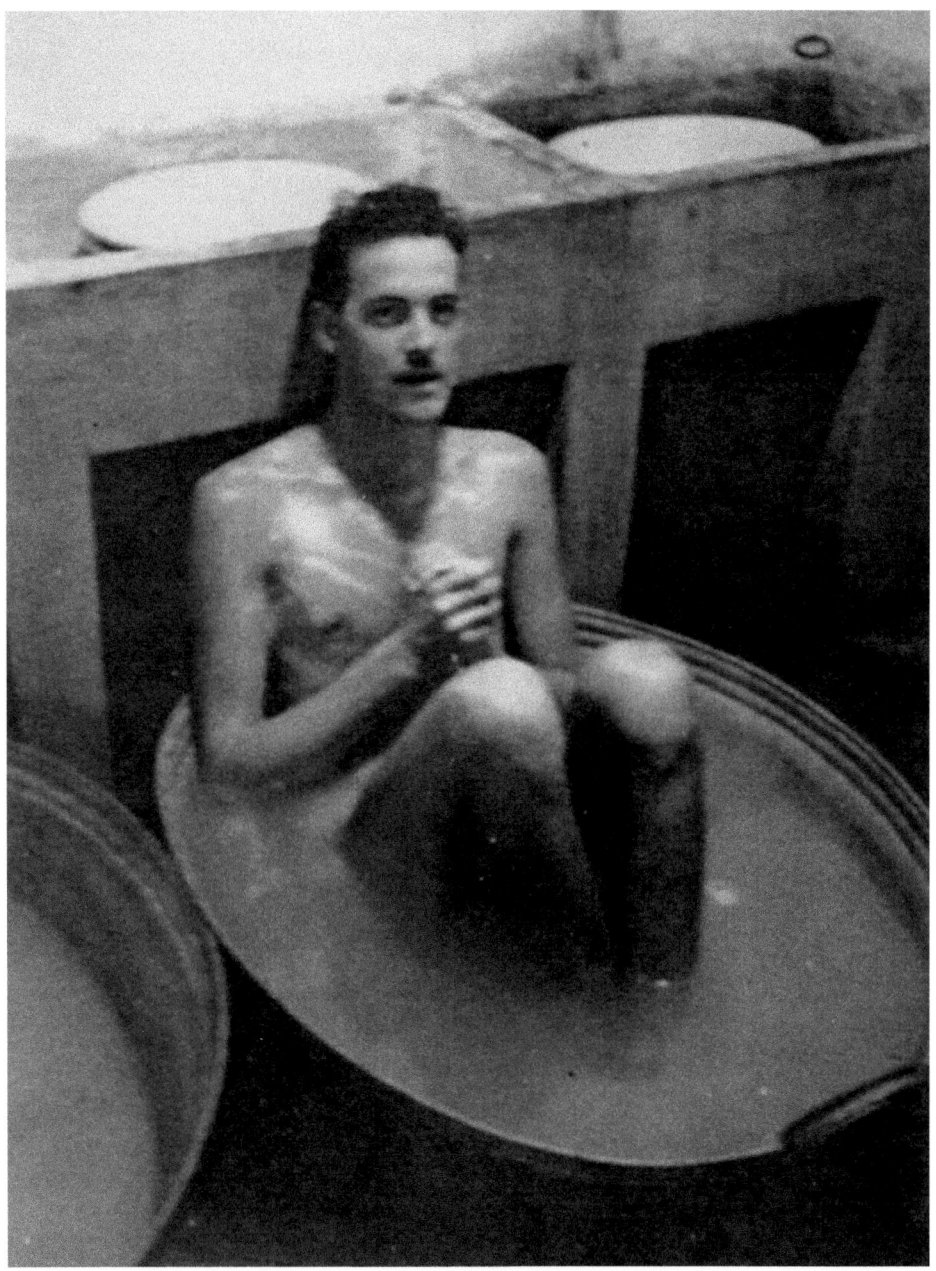

Ted in a zinc bath at Shallufa, on the Suez Canal, 1942. Author's collection courtesy of Herschell Reilley

from Malta, and didn't have a clue who we were, or where we were supposed to be going. Not wanting to return to the Far East, and feeling that we had slipped anonymously through the system, Herschell made a suggestion. 'We don't want to go back to the Far East. Here's our chance. We could quite easily take the plane and fly it all the way back to the UK.'

While the others contemplated this proposal, I responded with: 'Don't be bloody crazy, Hersch, I'm not spending the rest of the war in a guard camp in some Godforsaken place.'

Just then, an officer approached and announced: 'There will be a Mark 1 ready in a few days, for your onward posting to Ceylon.' So, our fate was sealed and the Far East it would be!

In the meantime we were to be at RAF Shallufa on the Suez Canal for a few days. At dinner in the mess that night, an American officer sat next to us. He was stationed at an American airbase on the Great Bitter Lake near Ismailia. He told Herschell of a Canadian, by the name of Charlie Harrison, who flew Flying Fortresses, and wondered if he knew him. He didn't. Herschell then turned to me. 'Ted, you ever been on an American station? We've got the time. What say we go and check out 'Cousin' Charlie?'

I replied, '*Ja*, man, why not?' We asked the American if we could get in on an inspection tour. 'Sure, no sweat,' he replied.

Permission obtained, we borrowed a Land Rover from the motor pool and I drove behind this American chap we had met. When we got there, Cousin Charlie was flying above the station. We looked skywards and at the height of about 20,000 feet we could see the dot of the plane repeatedly doing a figure of eight pattern in the same place. I eventually said, 'What the bloody hell is he doing up there?'

The Americans smiled. 'Why, he's making the ice-cream for supper!'

Herschell filled in the details:

> That night after clearing our throats of the desert dust in the mess, we adjourned to the dining room, where we were joined by 'Cousin' Charlie. We proceeded to eat an excellent meal, complete with blueberry pie and ice cream. ... So how does the ice cream wagon work? Well, the machine shop had made some 30-gallon stainless steel cans with suitable attachments so that they could be fitted onto the bomb rack. Then, when filled with ice cream mix, they were slung into the bomb bay of the Flying Fortress. To get maximum freeze, the bomb bay doors were left open. The ice cream manager knew just how much flying

DOING THE FLYING CHA-CHA-CHA

produced a good solid freeze. ... Yes, they did things a bit differently in the USAAF. We wondered just how far we would get if we made a similar suggestion to our RAF outfit.

Driving back, through Ismailia from the American base, we all noticed a very large yacht tied up at the wharf. Brownie said, 'I wonder who that belongs to.' I slowed the Land Rover and we all took a closer look and were aghast at the opulence of it.

Wilkie piped up, 'I was talking to a joker at the air station and he said that it belongs to King Farouk. He used to have it tied up at Alexandria, but he kept trying to escape to Italy. So the Navy had to go and bring him back and put the yacht at Ismailia. Now he can just sail up and down the canal. He's a weird boy, the king.'

On our leaving Shandur, the officer in charge of the base came across to say, 'I understand you are going to Ceylon. Good luck and stay out of trouble.' He shook our hands and we boarded for take-off to fly back to LG224 before finally leaving for the Ceylon trip. We took off on 1 November '42, now back in the cumbersome and laboured Beaufort Mark 1, and flew again, to Habbaniya, to Bahrein, and finally on to Sharjah again.

Before leaving Sharjah, my navigator, Hersch, asked, 'Where to now? I don't suppose you want to fly the long, five-hour haul straight to Karachi?'

I said, 'No, that's quite a lengthy trip, I think we should take a break at Jiwani in India, on the Persian border.'

'Good thinking,' he replied, 'I'll do the navigation plot right away.'

We took off and were no sooner airborne than one of the oil pressure gauges goes roomph and shows there's no oil pressure in one of the engines. So I very smartly put her in a turn, whistled round and landed again. This meant another couple of days in Sharjah, and hell, it was hot there, bloody hot!

Anyway, they fixed the oil gauge and we finally got away and flew to Jiwani. When we arrived over Jiwani we circled it twice, looking for the aerodrome, but there wasn't one. Hersch then spotted a long black line crossing an area, which appeared to be fairly flat, and pointed to me and shouted, 'Aerodrome'.

I shouted back, '*Ja*, it must be a landing area.'

I lined up and we landed. We were met by the CO, Mike Cheshire, in a lorry, and quite alarmed he said, 'Sorry chaps, we are only a signals station; we've got no fuel or mechanics or spares, we are not an operational aerodrome.'

'Oh, that's OK,' I assured him, 'we have plenty of fuel and don't need any mechanical assistance. We have only stopped to break the journey to Karachi, because we're a bit cheesed off with this flying backwards and forwards.'

ON LAUGHTER-SILVERED WINGS

Jiwani tented camp in an oasis on the desert runway, 1942. Author's collection by courtesy of Herschell Reilley

He transported us to what was literally a camp in an oasis. There were only six members of staff, all signalmen, camped on one side of the oasis. On the other side camped ten Arab labourers whose job was to run the oil truck and renew the oil line down the centre of the 'runway'. It was quite beautiful to find this lush little island of vegetation in the desert. The camp consisted of tents pitched under palm trees, and although it was pretty basic, there was lots of food, whisky and sleeping accommodation. We were given a large square tent to sleep in; it had four camp beds, each covered with a mosquito net. It was very hot.

Of this night, Herschell writes:

> The sides of the tent were rolled up for ventilation, and from our cots [camp beds] we could see the sky. My, they made stars big in that part of the world. I spent a good deal of the night just lying there and looking at them. The other thing that kept

DOING THE FLYING CHA-CHA-CHA

me awake was the donkeys. Each of them had a bell around its neck that tinkled as they moved around. Between the bells and the stars and the breeze through the date palms, all in all, it was one of the most pleasurable nights that we had ever passed. We all just lay there and absorbed the environment, each one gently falling asleep, and sleeping the sleep of the just. We were all up early the next morning, ready to fight the war.

We took the two-hour flight back to Karachi, luckily arriving in time for lunch at the airport's splendid restaurant. After our last experience of Karachi, we weren't keen to hang around there, but because the plane needed a service and an inspection we had to stay a few days. This time we became honorary members of the Gymkhana Club, which was a normal courtesy extended to visiting 'dignitaries' and was standard for all clubs in British India. The club was frightfully posh, had an indoor swimming pool, tennis courts, and a cricket pitch and polo field. We took full advantage of the swimming pool while we

Playing cards at Jiwani, 1942. From left: Hersch, Brownie and Ted. Author's collection by courtesy of Herschell Reilley

were there, and all agreed that roughing it in the tropics wasn't so bad after all!

Three days later, our aircraft was ready and we flew down the Indian coast to Bombay, where we stayed the night. We continued down to the Yelahanka Aerodrome at Bangalore, which is situated on the mountain range in the centre of the southern part of India. I had some generator problem in one of the engines on the flight down, so while that was dealt with we were held up there for four days. Bangalore is about 3,000 feet above sea level and has a much cooler climate than the jungle below it. It was a beautiful place, with lush vegetation and a large variety of flowering plants, and made for a very pleasant stay. We eventually left Bangalore on the last leg of what had been a very long journey. We flew across the Gulf of Manaar and finally landed at Ratmalana Aerodrome in Ceylon, on 9 November. Our journey had begun on 19 August from Malta; it had taken us two and a half months to finally arrive at our destination.

Chapter 16

Thumb Twiddling

*Always keep Ithaca in your mind
To arrive there is the ultimate goal
But do not hurry the voyage at all
It is better to let it last a full year ...*
C.P. Cavafy, from the poem *Ithaca*

After months of living on the bleak, rubble-heaped island of Malta, with the screaming and thunderous noise of air raids as our constant companion, the crew and I now found ourselves in quite a different world. From the air, Malta had looked like a tiny, scarred moonstone set in an ultramarine sea. In contrast, the island of Ceylon was like a huge faceted emerald, rimmed with white gold, and floating on liquid turquoise silk.

Beauforts flying over Ceylon. Picture taken from Ted's aircraft. Author's collection

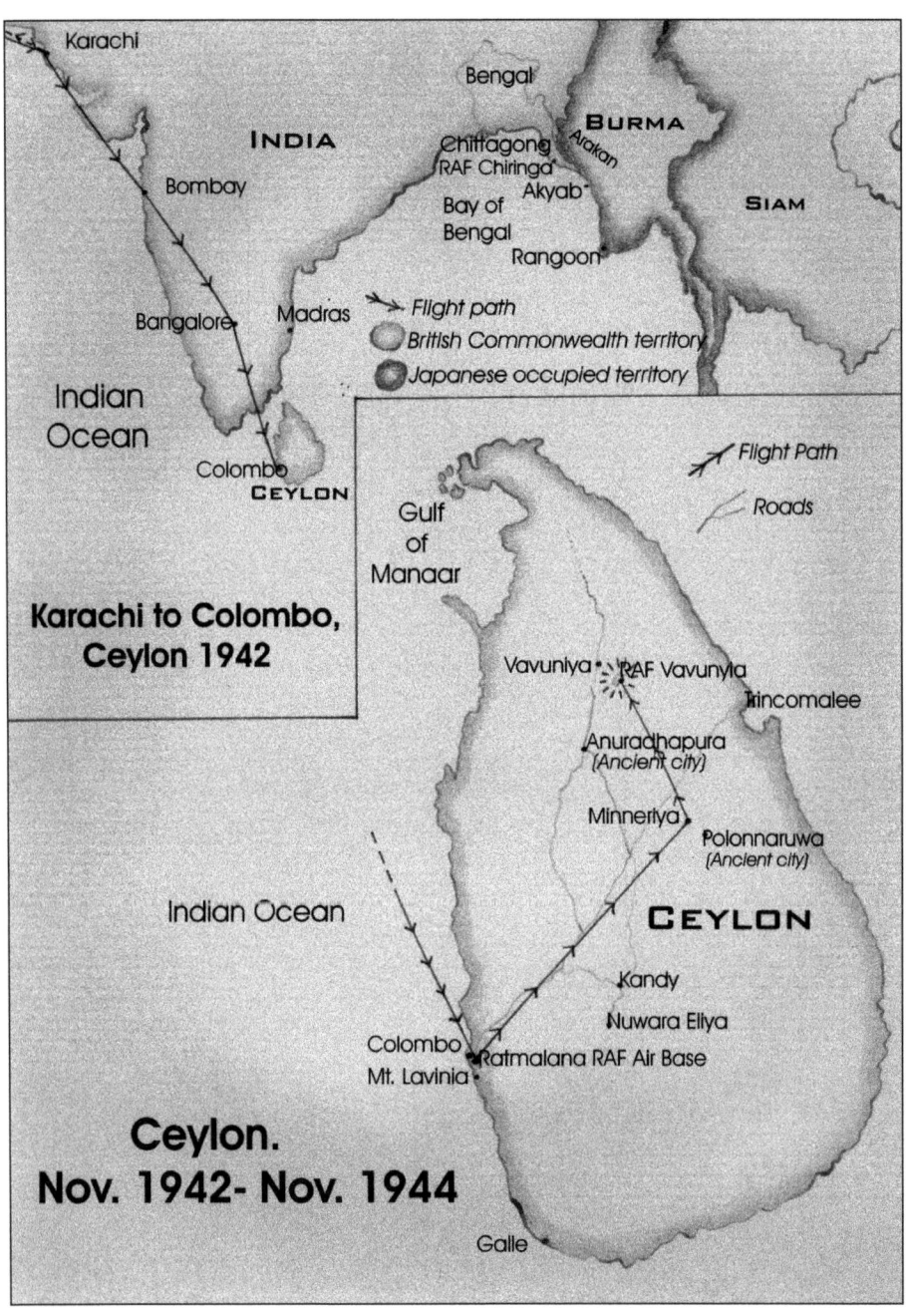

Route to Ceylon.

THUMB TWIDDLING

On arrival at Ratmalana Aerodrome in Ceylon, the plane needed repairs again, so while the mechanics were busy with it we took a seesawing rickshaw ride, in pairs, into Colombo. Unlike the war-torn Malta, and the chaotic confusion of Indian cities, Colombo spoke of order. Well maintained, elegant buildings stood on uncrowded tree-lined streets, where elegant turbaned Indian traffic officers assured the smooth flow of traffic. After a visit to the post office to send news home, we spent the rest of the afternoon exploring the town and docks.

That night we had a splendid meal at the Grand Oriental Hotel, which like the Karachi Airport restaurant was opulent and luxurious. Malta's bully-beef fare and the feeling that we would never eat our fill again were forgotten. Upstairs, we slept in palatial comfort to be woken with tea and pancakes in the morning. After this, we had a full English breakfast in the dining room before leaving all this luxury to get back into the plane and reality.

Eight months before our arrival in Ceylon, the Japanese forces had attacked Trincomalee, the naval base on the east side of Ceylon, on Good Friday 1942, and on Easter Sunday had attacked Colombo. Both attacks were carried out with little or no resistance from the British forces on the island. They were totally unprepared for it. Hence, fearing another attack on the next religious holiday, which was Christmas, they had called for more air defence for the island. So here we finally were – at our original and planned destination, after being rudely commandeered en route into some fierce action on Malta – waiting for another Japanese onslaught.

At Colombo headquarters we were elated to be handed, in bundles, six months' accumulation of mail from our loved ones at home. In spite of not having received any mail since leaving England, I had continued, with hopeful expectations, to write to Bee whenever I got the chance. I was apprehensive about what her letters held for me but my fears were immediately dispelled by the sheer quantity I received from her that day. Her reciprocated feelings gave me the courage to write back with a marriage proposal, which she accepted. My mother in Johannesburg bought and posted an engagement ring to Bee in Bulawayo. We were thus engaged by proxy, in early 1943. Bee then sent me a gold signet ring engraved with my initials, and inscribed with love on the inside. Although not one for wearing men's jewellery, I wore it on my little finger in Ceylon, as a symbol of our commitment to each other. During the next two years there was frequent correspondence between us ... or so I thought.

The crew and I were immediately transferred from 217 Squadron, which in Ceylon had changed over to flying Lockheed Hudsons, to 22 Squadron, which was still on the old Beauforts. To join 22 Squadron we flew inland from Ratmalana to the airbase at Minneriya in the jungle, north-east of Colombo, and arrived there on 12 November 1942, landing on the single runway cut out of the jungle. Soon afterwards, I was promoted to captain and became the deputy flight commander of 22 Squadron's B Flight. Our activities consisted of anti-sub patrols and escorting ships that were travelling through our area from Aden to Australia. We were really just twiddling our thumbs while waiting for the Japanese to attack the island again, which they never did. It was a terrible waste of time. I suppose that someone had to be in Ceylon because at the time the Japanese were having tremendous success in Burma and had taken it right up to the Indian border.

After the newly swelled ranks of the squadron had shaken themselves out and settled into their jungle home at Minneriya, a third of them went down with malaria. Hell! It was incredible. The place was rife with mosquitoes.

Vavuniya airstrip from the air. Author's collection by courtesy of Herschell Reilley

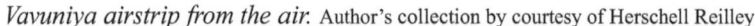

THUMB TWIDDLING

A Beaufort taking off from Vavuniya airstrip, 1942. Author's collection by courtesy of Herschell Reilley

Never mind the war, in Ceylon mosquitoes became our foremost enemy! For some reason I managed to evade the dastardly infection.

In February 1943, the whole squadron was transferred to Vavuniya, which also had a single narrow runway hemmed in on all sides by the jungle, and was about 50 miles from Minneriya. Vavuniya was a much better built and maintained station and was occupied by about 600 servicemen. The runway had been constructed by using the only source of power available – the wonderful Indian elephant.

Eddie Whiston wrote to me in 1997:

> It was bumpy and wavy, not quite straight. There were dispersal points or parking places lining the runway and on either side the trees of the forest dwarfed you. If an aircraft swung right or left by just a few degrees there was an inevitable crash. During my time there, the airfield was three times condemned as unsuitable for Beaufighters. However, this made no difference and we still went on flying.
>
> Hygiene, sanitary and food arrangements were all provided to a very high standard. Tracks were made using elephants, and ran between the dispersal areas and the various workshops and offices. These were wide enough for a jeep or small lorry and they were all given the names of London Streets, e.g. Regent

ON LAUGHTER-SILVERED WINGS

Street etc. ... Of course, strict anti-malaria procedures were practised, such as spraying working areas with DDT and daily doses of anti-malaria medicine.

Dad takes up the story again ...

We lived in long low bungalows called bashas, which, apart from the concrete floor, were built entirely from coconut trees. Each hut was divided into eight units, and the whole framework was constructed with coconut poles and tied together with coconut fibre rope. The roof covering was made from coconut leaf fronds split down the middle, the half-fronds fringed and then plaited together into a mat. The mats were then tied onto the frame and overlapped to produce a waterproof surface. The roof cap along the top length of the hut was made of whole unfringed fronds. These structures were very practical in the extreme tropical heat of Ceylon, and were very comfortable. The fact that they blended so well into the jungle environment meant that they attracted all sorts of unwanted visitors of the wildlife variety. Monkeys were often in and out, causing havoc, leopards prowled around at night and then there were the snakes!

In the heavy, humid jungle heat it became routine to take a siesta after lunch. One afternoon while we were all lying on our beds, there was the sudden outburst of: 'Jesus Christ!' Everyone bolted up off their beds and found one chap pointing to the main beam of the hut, at an 8-foot snake as thick as a man's wrist lying along the length of the pole. No one was going to find out if it was harmful or not; it was too close for comfort, suspended as it was over our heads. With much machismo and two poles, it was dislodged from its resting place and dropped to the floor, while men shouted blasphemous exclamations and leaped in all directions. Seemingly unperturbed, it slithered out of the hut with great speed, leaving us all in a variety of defensive postures, some standing on beds, some armed with whatever they could find and some, legs in the air, hanging onto the roof beams. We must have looked like a bunch of bloody fools.

The toilet facilities at Vavuniya were rudimentary but functional. They were built about 100 feet from the sleeping quarters and consisted of a long rectangular hole dug out of the ground. It was about 10 foot deep and about 4 foot wide, a bit like a deep and narrow trench. Over this trench was suspended, at about 15 inches above ground level, a thick plank, with eight appropriately shaped holes cut out of it at regular intervals. These holes were divided from each other by coconut matting walls, which created a flimsy cubicle to offer a modicum of privacy. However, below the plank, there were no such divisions along the inside of the trench. For obvious reasons of hygiene and pest control, a mixture of gasoline and fuel oil in large quantities was poured into the trench every couple of days.

THUMB TWIDDLING

With much amusement, Herschell Reilley, Dad's navigator, wrote to me about a drama that issued forth from this facility:

> While Ted was not a principal player in the following drama, he was most certainly involved once … when one entered these facilities, the last thing on one's mind was the possibility of disaster. There was nothing better to do than contemplate happy times – or have a cigarette … it is not difficult to anticipate the result. In this instance the result was positive, instantaneous, and quite violent. The lighted match hit the gas, the gas went boom, and three people were virtually catapulted into the jungle. The flame ignited the coconut matting and it was well lit very quickly. The three turned and looked at the conflagration in something akin to wonder. A short time later, a very short time, the results of the initial explosion became felt. Each put their hands on the sore part of their anatomies, and withdrew in a hurry. Someone ran to the mess and called the doctor and the fire station. They arrived a few minutes later, and the fire was quickly doused. The doctor's advice to those who had been wounded was, eat standing up and sleep on your stomachs. As a matter of fact, two of the three were feeling well enough to give the perpetrator a liberal sprinkling of four-letter words. Yes, your dad was in the next cubicle!

Towards the end of 1943, Dad was feeling that he had been kicking his heels for most of the year in the jungle at Vavuniya, and he was pleased to be given command of a detachment of Beauforts. Pilots and crews would be operating from Bombay.

Chapter 17

Bombay and Biding Time

We flew over to Bombay with these Beauforts to operate out on anti-sub patrols to assure safe passage for the convoys of our troopships that were coming into India. On one of our patrols out of Bombay, one of our aircraft piloted by Harry Haffrey went missing. I sent up five aircraft to fly line abreast to search the area where it had gone down. It was a huge area and nothing was seen of these chaps. It was a terrible moment for me and I was preparing to write to the families of those missing. These were awfully difficult letters that commanders had to write and I never got used to them. Anyway, about four days later, I got a phone call to say that there were some chaps in hospital who had crashed into the sea and been picked up by an Arab dhow. Thank God! It was my missing crew.

According to Harry, 'the sea was so glassy that I had difficulty judging where the water was, and I flew straight into it.' I quite believed him because the water around Bombay was often like a mirror. As the plane was sinking, the crew, all luckily wearing their Mae Wests, had tried to release the dinghy without success. Quite miraculously and against all mechanical logic, one wheel of the aircraft popped up to the surface in the midst of the men.[1] Without this float to hang onto they would not have survived the night, let alone a few days in the sea. They were seriously sunburnt from being exposed for so long but we were so pleased to have them back again and there was great rejoicing when they returned from the hospital. I asked Harry, 'Didn't you see the Beauforts searching for you? How did they miss you?'

Harry replied: 'We saw them, but when they flew over they were just too far apart to see our bobbing heads between them, and I suppose they were looking for a dinghy.'

And, oh Lord, every fortnight I also had to be the paymaster for these chaps. I was very young, twenty-three, and having all this responsibility worried me a bit. Anyway, the first time I went in to collect the money from the pay office, I was handed, I don't know, about 20,000 rupees, and a rupee was the equivalent of about one and six at the time. Because I had all this money, and being very conscientious, I thought I had better check it by counting it. Well, by the time I got to a thousand rupees I gave up and put it all in the bag and took it back to the aerodrome. Fortunately, it was correct and none of it was missing.

After five weeks in Bombay we were relieved and sent back to Ceylon. On the way back we had to refuel at Bangalore. Well, this wasn't a very satisfactory state of affairs, because the fuel there was stored in old 44-gallon drums. These had to be lugged onto the wing and the fuel poured through a funnel. There were no shammy leathers; it was an ordinary funnel, and the drums had been stored Lord knows where. ... Anyway, it took a couple of hours to get all the fuel into the tanks and we set off again. We crossed the Gulf of Manaar, the sea between India and Ceylon, and I got a splutter in one of the engines, and then a cough from the other, then both engines were spluttering and coughing. I realized then that we had water in the bloody fuel.

We still had about 46 miles to go to get to our aerodrome, and we were flying across thick, dense jungle. Because of the clouds we were flying fairly low at about 800 to 1,000 feet, with the engines going hic-pup-pupbrrrr-pup-pupbrrr, and this was a very unhappy state of affairs. I was just sweating it out, well aware that if they cut or died on us we would be in big trouble. We were far too low to bail out, we couldn't fly above the clouds because we had no navigation aids at all, and attempting a forced landing in the thick jungle forest wasn't an option. Thank God, we managed to splutter and stutter into Vavuniya Aerodrome safely and, sure enough, when the tanks were cleaned out there was water in the fuel.

We all settled back into our jungle routine again just before Christmas of 1943.

Any leave we got was spent at the coast in Colombo or up in the hills at Nuwara Eliya. Both places offered relief from the stifling humid heat of Vavuniya. Nuwara Eliya, where all the tea estates were, is a village set on a plateau about 6,000 feet above sea level. From the village one could climb a peak called Mount Pedro and experience the most glorious panoramic view in all directions.

Herschell describes it in a letter:

> From the top of Mount Pedro one could see up the east coast almost as far as Trincomalee Harbour. To the south one could see the sweep of coastline. To the west, one could see Galle, a small coastal town. ... Ceylon is 400 miles long, and to the north both the east and west shorelines on either side disappeared into the horizon. ... The surrounding mountain slopes were covered

with tea plantations. ... At any given time, areas were covered with tea pluckers – girls with large woven baskets on their backs. ... The girls would pluck a handful of new leaves off the shoots and throw them over their shoulder into the basket.

Other sightseeing trips Dad and his crew took were to the ancient ruins of Anuradhapura and the medieval ruins at Polonnaruwa, which lies in the centre of the island. Anuradhapura was established in the fourth century BC and served as the capital of the Anuradhapura kingdom until the eleventh century AD, when the capital was moved to Polonnaruwa. Both places are now Buddhist World Heritage Sites.

Among Dad's photographs is one of him and his crew standing in front of,

Ted and his crew at the ancient town of Polonnaruwa, 1943, Ted standing bottom left. Author's collection

BOMBAY AND BIDING TIME

Ted riding an elephant in Ceylon, 1943. Author's collection

and positively dwarfed by, the colossal statues of Buddha at Polonnaruwa. The statues, depicting the final moments of Buddha's life before he entered the state of enlightenment, are carved out of one solid granite outcrop. In the photo only two of the four statues are visible. Other photographs show Dad and his crew enjoying rides on the hard working Indian elephants.

When I was given leave I spent a lot of time in Colombo because I had met the Marches, a wonderfully hospitable couple who lived there. Eric and Doreen March offered me a home from home. Eric had welcomed me with, 'You are very welcome, you don't have to bring anything or pay anything as a serviceman; the only thing I ask is that you tip the servants because it's extra work for them.' This was very fair and they were fantastic to me. Eric was a government official in the Forestry Department and they had a huge house in Colombo. They rented out rooms on the ground floor to what they called 'PGs', or paying guests. Anyway, one of the suites was rented to the Hall family.

Mr Hall was a customs official in Colombo, and this I thought was quite amusing, because he didn't drink whisky and at the mess in Vavuniya we got a ration of one bottle of whisky, two quarts of beer and two bottles of wine a month, which wasn't a helluva lot. This shortage led us to discover that while

The 'boys' swimming up the coast from Colombo. Author's collection by courtesy of Herschell Reilley

most Indian alcohol was vile, the Indian gin wasn't too bad, albeit a bit scented. As we had the occasional exercise to fly over to Trichinopoly, we would take two empty demijohns with us and fill them with Indian gin and smuggle them back to Vavuniya. Then we would take the empty gin bottles from the mess and fill them with gin from the demijohns. Whenever I went to Colombo on leave I would gaily go and swap this gin for whisky with Henry Hall, the customs official.

Dad also met some other friends – Peter and Jackie, and Jinx – in Colombo. They would go out and about together. Jinx Glennie-Carr, wrote this to me:

> Jackie and I were Wrens in Colombo, stationed at HMS *Bermuda*, the Fleet Air Arm station. Jackie and Peter were married at the time. Peter met Ted, who was attached to the RAF at Vavuniya, at an aerodrome in the jungle in the north of

BOMBAY AND BIDING TIME

Ceylon. Peter, a captain in the Royal Artillery, had been on a jungle course up there. So the next time Ted came down to Colombo, Peter and Jackie brought him round to collect me to go to some party. Peter's car, a small open sports job, was almost impossible for Ted to get into – there was no way he could be folded up to sit in the back. However, Ted and I would sit on the folded-up hood with our feet on the back seat. Ted taught us to sing *Sarie Marais*, which we would do at the top of our voices as Peter drove us through Colombo – to the amazement of the Singhalese! Whenever he was down from Vavuniya we included him in all our parties: nightclubbing at the Silver Fawn, dining and dancing at the Galle Face Hotel and our swimming club at the beach at Mount Lavinia. We did work as well! I was a wireless operator and Jackie was a cine gun assessor.

Dad recalls …

I found Colombo such an interesting and exotic place. Every second shop was a little jewellery shop crammed with precious and semi-precious stones like the star sapphire, which Ceylon is famous for, and some were as big as pigeons' eggs. There were also real zircons, rubies, emeralds, sapphires, and an array of multicoloured semi-precious stones. It was fascinating to watch the stone merchant turn and polish the cabochon stones. He had a piece of string tied to his big toe, which was attached at the other end to a treadle. As he moved his foot to and fro the treadle would work the polishing lathe. I loved watching this process, perhaps because my childhood was spent watching my father at work; I have always been intrigued by artists and craftsmen at work. While I was in Ceylon I bought quite a few little stones to bring home for Bee: zircons, pink tourmaline, an oblong-cut pale pink ruby, a tiny star sapphire, and an aquamarine and zircon pendant and ring set. In Ceylon, these stones were mined mostly from holes in the ground with two chaps digging at the bottom with poles. Whatever they found was sent up to the surface in a bucket on a rope.

For some time, in Vavuniya, we had had a Beaufighter squadron with us as night fighters, and I had got on very well with the chaps, the flight commanders and the CO. By this time I had been promoted to major, which was the equivalent of a squadron leader and flight commander in the RAF. By now our squadron had been in Ceylon for about sixteen months on the Beauforts and, in mid-1944, the RAF finally decided to change us over to the Beaufighters. We were all thrilled because the Beaufighter was a beautiful aeroplane. It was wonderful doing torpedo-dropping exercises with the Navy, and exercise attacks on the harbours were all thoroughly more enjoyable with this more efficient aircraft.

Towards the end of 1944, a signal came through informing me that I had been promoted from major to lieutenant colonel, and was to take command of 211 Squadron in Burma. 211 Squadron was a Beaufighter squadron based at Chiringa, which was then in India, near the border with Burma. I was overjoyed because after two bloody years of fiddling around in Ceylon, I was at last back in action.

Chapter 18

A Flamer – As Fate Would Have It!

Ithaca has given you a beautiful voyage.
Without her you would not have taken the road.
But she has nothing more to give you. ...
C.P. Cavafy, from the poem *Ithaca*

Within days of this notification Dad was to take up his post as squadron leader of 211 Squadron in Burma. He continues with his story ...

The evening before I left Vavuniya, we had a farewell party in the mess and a big pack-up. The next morning was 1 November 1944, and before leaving the station I went out in a Beaufighter to do a bit of rocket practice because we were going to use them in Burma. I remember noticing it was called V for Victory, the irony of which would only hit me later. After the practice, I landed, had lunch and said cheerio to the chaps.

An Australian pilot called Boyd Gerny had been assigned to fly me to Colombo to catch the transport service to Calcutta, and from there across to Burma to join 211 Squadron at a place called Chiringa, in the Arakan.

We got into the aircraft at about two in the afternoon, and hell it was hot! It used to be so hot there that, if I was flying, I always wore my gloves because there were no hangars or covering for the aircraft at Vavuniya and the metal controls got so hot that they would actually burn your hands.

In the aircraft I was sitting behind the pilot, Boyd Gerny, between the wings in the fuselage. The Beaufighter has a very narrow fuselage, so the crew cannot sit side by side. The pilot sat in front and the navigator, Warrant Officer J.B. McCouat, sat about 4 yards behind him. In this instance I was sitting on the spar between them and I can remember it so clearly. I was wearing, my leather flying helmet so that I could hear the intercom. The microphone for the intercom was set into the oxygen mask. I had put mine on.

We set off and taxied out to the end of the runway and prepared for take-off. As we were taking off, the tail had just come up off the ground when she started to swing. She swung to the right, then she swung to the left and then she swung right off the runway. Now out of control, we careered past the control tower, crashed into a stationary four-wheeled mechanics' workshop, and then crashed straight into the parked Spitfires in the dispersal area. As we hit one of the Spitfires there was an almighty … CRASH … BANG … VVVWhoomph! … The whole bloody lot just exploded into a gigantic moving ball of flame … and as the plane continued hurtling forward I saw the trees looming up, and as we hit them and came to a standstill, the left perspex window shattered. And we were in the middle of this inferno.

I remember two sheets of flame, like welding torches, shooting across the fuselage from each side just in front of me and catching Boyd Gerny, who like a fool had pulled off his helmet. He then whistled through the shattered window and disappeared. I followed him and, as I got out onto the wing in the midst of all these flames, my interlocking connection for the microphone, which is a fairly tough cable, pulled me back. I turned around in a frenzy, grabbed it with both hands, and with the strength only adrenaline enables, I broke the cable. The Beaufighter was still high on its wheels and as the cable broke I sailed backwards off the wing, landing flat on my back amongst all the flames. I remember thinking to myself, 'Oh you bloody fool, you've knocked yourself out; now you're finished' … and I passed out.

As I came to and rolled over onto my right side the petrol tank in the wing above me exploded and split open, spewing petrol over and around me. It hit the ground with an almighty VWH…OOMPH, and burst into flames. I don't know how I got up, but I ran, and ran, and ran … out of the hellfire. … When I stopped running I remember saying to myself, 'My God, you're lucky you're not even burnt!' I sat down and only then did I look at myself. And, oh God! What a mess! The full length of both my legs, and my arms down to my fingertips, was a mass of charred grey strips of flesh. It looked as though someone had stuck grey ribbons all over me. The flesh was just hanging off all over the place. My skin had been burnt off. I was wearing a short-sleeved shirt and shorts with my socks pushed down to my ankles. I hadn't yet seen my face and still had on my leather helmet. I remember clearly being completely and utterly still in control of myself physically, but the shock is very hard to describe; you just don't know what to do.

I'm not sure whether it was from traumatic shock, but I sat there within 20 meters of the blazing inferno of this Spitfire and Beaufighter, which between them had some 800 gallons of fuel going up in flames. I can't describe my feelings – my mind seems to have been in a suspended state of neutrality. I didn't know whether to get up and run, or whether to cry out, or whether to

A FLAMER – AS FATE WOULD HAVE IT!

The Inferno: blazing Spitfire and Beaufort, Vavuniya, 1 November 1944. Author's collection

Vavuniya Base, with the burning wreckage in the background, 1 November 1944. Author's collection

scream. I felt no pain, I think, because the adrenaline that flows obviates anything else. I couldn't feel anything, and with third-degree burns, the skin and nerves are burnt away. I saw Boyd Gerny a little way away from me also just sitting there. I don't know how long I sat there; time seemed suspended, until suddenly a tyre burst, exploding with a very sharp ... BANG!

Instantaneously, we both leapt to our feet and started bolting, just running, and running to God knows where. ... I ran around one of the bashas, and there in front of me were four or five squadron blokes who were running towards the crash. They shouted, 'My God, there's Ted!' With that, my legs buckled under me and I collapsed. (I think it was knowing that help was at hand that did it. Had I not seen them I think I would have kept on running.) Boyd Gerny had torn down the runway in the opposite direction and they had to actually rugby-tackle him to get him to stop running. He was in a severe state of shock.

They eventually got both of us into the airbase ambulance and into the local sick quarters on the Vavuniya base. By the time they got me out of the ambulance my fingers were starting to curl up like animal claws; I had no control over this. In sick quarters, our wounds were wrapped in rudimentary coverings and we were each put onto a drip. The ring that Bee had sent me had to be cut away and extracted from my finger, and although I kept it, I never wore it again. There we waited for some air transport to get us to the hospital at Colombo.

At this stage I was conscious, not yet in pain, and all I could think about was losing my posting to command 211 Squadron in Burma. I remember insisting and pleading to the staff, 'I don't want to miss my posting to Burma, please just tell them to keep my posting,' assuring them, 'I will be fine in a week or so,' and kept repeating, 'I won't be long, I won't be long.' I lay there fully conscious for the rest of the afternoon. The crash had occurred at 2.00 pm and at 5.00 pm a Dakota arrived to fly us out to hospital. I remember not being keen to fly at that particular time. On the flight I became aware that three of us had been burnt in the crash. I assumed the third person was the navigator, McCouat, who was with us in the plane. However, I had yet to learn of the full extent of our terrible crash.

I even noticed that the pilot of our mercy flight, Major Frank Robertson, was wearing South African Air Force uniform, and I chatted to him on the flight. I remember landing at Ratmalana Aerodrome, outside Colombo. I was aware that Group Captain Paul was there to meet us and heard him saying, 'Ted, I will ensure that whatever you need, we will get for you.' Although I was of a much lower rank than he was, we were great friends and whenever I had flown into Ratmalana, I would pop in for a chat and a cup of tea. He really was a terrific chap. We were taken to the 35th British General Hospital.

A FLAMER – AS FATE WOULD HAVE IT!

There we were put into a general ward, examined again, told we would be monitored through the night and then heavily sedated with morphine.

In the morning I woke up mummified. My head, my hands, my arms and my legs were all strapped up in bandages. I couldn't hold anything and I learned that the right side of my face had been burnt. Although I had been heavily sedated, I can quite clearly remember everything that was going on around me. One of the air mechanics came up to see me and said, 'Ted, I am so sorry this has happened, is there anything I can do for you?'

I replied thankfully, 'Just light me a cigarette.' This he did, and I was able to smoke it with him holding it to my mouth.

It was only at this time that I learned the details of the crash. I had thought the third person with us was McCouat, and was hoping he had got out of the plane, but tragically he had died in the crash. The other person with Boyd Gerny and me in hospital was Corporal Edwards, one of the two wireless mechanics who had been working in the mechanics trailer on the runway when we had crashed into it. Because of the very hot weather that day, he had taken his shirt off and had only been wearing his shorts, and therefore was very badly burnt and in an extremely critical condition. Very sadly, he died in hospital six days after the crash. The other mechanic, LAC Reilly, had been killed in the impact.

The next day we were moved into a special ward and each given our own nurse. The nurse who took care of me was called Penny Clarke, and I couldn't help thinking it a strange coincidence as I was engaged to a girl who was also a Clarke. I was still conscious and was aware of talking to Penny that morning. At about 10.00 am, about twenty hours after the crash, they decided to put me in a saline bath, which was the only treatment for burn victims at the time. I could still manage, with help, to walk to the bathroom. When I was in the bath and well soaked, Penny started to take off my dressings. As my wounds were still fresh and moist the pain was bearable, but I suddenly went into convulsions. Uncontrollable shaking and jerking was followed by total physical collapse, my body just fell limp and powerless – very distressing, as I was somehow still conscious.

Penny called an orderly and they tried to get me out of the bath, but it was walled in on three sides, and after battling with the dead weight of my body, they realized that they wouldn't manage. More orderlies were called and eventually they managed to manoeuvre a sheet under me and finally got me back onto the bed, by which stage I was losing consciousness. For the next three days I remained unconscious and didn't know what was happening. Apparently, they continued the saline bath treatments – after the first bath fiasco these were now carried out more easily in a freestanding bath next to each of our beds. An urgent request had been sent to Group Captain Paul for these, and he had in no time at all sorted them out for us. Penny told me afterwards that she really believed I was dying.

It was at this time during my Dad's crash ordeal that Eddie Whiston of the RAF Personnel in Colombo visited his bedside. In a letter to me, he writes:

> The re-emergence of that terrible crash is quite traumatic for me. The hospital in Colombo was a Victorian-type building and probably among the best in hospital treatment in the whole of South East Asia Command. The room your father was in was private to him, large and airy. The nursing staff was made up of white European girls and local Indian girls. A large fan circulated the air and the bed was surrounded by a mosquito net. When I was there your father was lying in a saline bath, which stood next to his bed. He was apparently too badly burnt to bandage up (dressings were removed in the bath) and I think at this time they were just relying on the saline bath treatment.
>
> I was there to take written notes of the questions to be put to him by the high-ranking officer I was with, whose name I don't know. He put the first question to Ted, which probably was, 'What happened?' Ted's lips slowly formed an 'O' but no words came from his mouth. At this point I was overcome by nausea from the dreadful sight and smell and had to excuse myself. I dashed to the toilet, thinking to myself that it might have been me lying there as I had flown myself down from Vavuniya to Ratmalana (Colombo) that morning. When I returned to your father's bedside the officer said, 'We are not going to get anything here,' which I thoroughly agreed with and we departed. Back at the base I told the CO, and everyone who asked, that Ted was dying. From that point on and for years later I assumed he was dead. It was only from a casual and chance remark from Roy Nesbit [in 1992] that I discovered he had, after all, survived! I couldn't believe it, but having read the account he gave of the horrifying circumstances I am forced to realize what a fantastic will to live he must have had.

After these three days of delirium Dad's condition started to improve and he regained full consciousness. It was then that he had a visit from the CO

A FLAMER – AS FATE WOULD HAVE IT!

at Vavuniya, Johnny Lander, who said, 'Look Ted, I have to give you this letter from your fiancé, but please disregard its contents because she has asked me not to give it to you.' In this letter Bee was calling off the engagement. She wrote: 'You have not written to me for many months and have obviously lost interest, so I am breaking off our engagement.' She had already posted the letter when she received the shocking news, in a telegram, of Dad's crash and injuries. Not to add to his grief, she had immediately telegrammed CO Lander to stop the letter from reaching Dad. Dad couldn't understand why she hadn't received the many and frequent letters he had written to her. (It was discovered that, when he addressed them, his '4' looked like an 'A' and all the letters had gone to 3A Rhodes Street in Bulawayo, instead of 34 Rhodes Street. She eventually received, like he had the year before, a large batch of them, and the romance continued to flourish.)

Now, starting to regain all his senses, Dad was put again into a saline bath. This time, he says, 'My God! I nearly hit the roof. It was unbearable. I shouted that something was wrong and gave everyone hell.' The sister came and said, 'You are just making a noise, keep quiet.' He insisted, 'No I'm not, it's giving me hell!' She turned to the orderly and asked, 'How much salt did you put in?' He told her, and she said quite nonchalantly, 'Oh dear, that's double the amount.' And then had a giggle. Dad wasn't amused and gave them a tongue-lashing.

Dad continues ...

Dressing our wounds was another terrible ordeal, the smell of the flesh going off was dreadful and when the nerves started to grow and heal, the pain was excruciating. The bandages would stick to the wounds and it was absolute torture when they started to take them off. They tried ripping them off quickly and ... oh ... the screams were something to be heard. Eventually they decided that they must give us gas to get them off and this was wonderful. I would drift away, I didn't feel a thing and then very gently and peacefully wake up and the dressings were off.

> **Daring Airman Injured**
>
> POTCHEFSTROOM, Saturday. —Major E. T. Strever, D.F.C., S.A.A.F., who has been injured in a flying accident in the South-East Asia area, accomplished the feat, early in the war, of landing a foreign aircraft on Malta when that island was being badly blitzed (writes the "Sunday Times" correspondent).
>
> Major Strever was forced to land in hostile territory and was taken first to Greece as a prisoner. While he and other prisoners were being flown to Italy they overcame their guards and Major Strever took over the controls, thus capturing the aircraft intact.
>
> He was awarded the Distinguished Flying Cross for daringly dropping a torpedo on an enemy battleship in the face of violent anti-aircraft opposition. On another occasion he and his crew scored a direct hit on an enemy merchantman in the Mediterranean.
>
> He is a son of Air-Sergeant W. H. Strever and Mrs. Strever, of Potchefstroom.

Sunday Times *newspaper cutting from 12 November 1944, South Africa.* Author's collection

ON LAUGHTER-SILVERED WINGS

One morning, there was a lighter moment to the process of doing our dressings. They had finished doing mine and had started on Boyd Gerny. However, I think he was becoming immune to the gas and they couldn't get him under. Every time they gave him some gas and tried to get his dressings off he would scream. So the doctor thought the bloody gas machine wasn't working, so decides to test it out. He sat on my bed, put it to his nose, turned it on, and promptly keeled over onto the floor, out cold ... out for the count. Hell! We had a helluva laugh about that.

Of course, all this time I was having to be fed and if I wanted a cigarette someone would have to hold it for me, so I went off smoking for a while. We were terribly well looked after. Doreen March, Jinx and Jackie came to visit me and we were treated like VIPs.

My dad's logbook entry of this terrible crash, recorded under 'Results and Remarks', reads: 'To Rat. (Ratmalana) on Posting, Swung on take-off – A Flamer! McCouat killed'.

This was the first fatal accident that 22 Squadron had had in almost two years.

Of the steps taken by the airbase after the accident, Eddie Whiston writes:

> After the crash we were forbidden to fly in short trousers and instructed to wear full-length flying suits at all times, so as to keep as much of the skin covered as possible, to limit the effects of fire. The mechanics, however, worked in the open dispersal areas in the lightest possible clothing because of the heat, so if they were sprayed with burning petrol they were simply fried alive.

Dad continues ...

After about six weeks I was healing up well, but Boyd Gerny had to have skin grafts because he had taken his helmet off and his ears were very badly burnt. Eventually, I was able to get on my feet and walk around a little, and they suggested that I go out on a few weeks' convalescence. The doctor asked, 'Can't you go home?' I said, 'Well, I don't know but I could ask at headquarters.'

I was taken in the ambulance and placed in a wheelchair and wheeled into the senior air staff officer, and he said, 'The AOC wants to see you.' So I went in to see the old boy, Air Commodore Derstan, 'Dusty Derstan' we called

A FLAMER – AS FATE WOULD HAVE IT!

him. And, oh, what a wonderful AOC he was, absolutely super. He made a fuss of me and asked, 'What can we do for you?'

I replied, 'Well, Sir, the doctor suggested that I should go home for a few weeks.'

He said, 'Well Ted, I can't officially give you leave outside of the command, but I'll tell you what I'll do. I'll make you an official courier.'

The CO gave Dad a letter to General 'Boetie' Venter, who was then chief of the South African Air Force and based in East Africa.

The cartoon taped on the inside of Ted's log book: 'Yes, it was a bit of luck – not even singed.'

Chapter 19

The Soft Touch of Home

*And if you find her poor
Ithaca has not defrauded you with the
great wisdom you have gained ...*
C.P. Cavafy, from the poem *Ithaca*

Just before Christmas 1944, I was handed some mail and the 'official' letter to General Venter in Nairobi, which read: 'This is one of your chaps who is under our command. He has been through a rough time. Look after him and do everything you can to help him on his way back to South Africa.' I was sent to Koggala, which is about 50 miles south of Colombo on the coast, to catch the Catalina courier service, which ran across the Indian Ocean between Ceylon and East Africa. I was still weak and had to have the dressings on my right leg attended to as it hadn't yet healed well. In spite of my condition it was an interesting trip across the Indian Ocean. The first part of the journey was a seven-hour flight to the tiny island of Addu Atoll, where we stayed overnight. The next day we flew the ten- to twelve-hour trip to the Seychelles for another night's stopover. From there we flew to Mombasa, where I caught the train to Nairobi.

At Nairobi I went to Air Force headquarters and gave them the mail I had from Ceylon. I knew that in order to connect with the shuttle service to South Africa, I had to get transport to a place called Kisumu, which ran four or five flights a day. So I asked to see someone at Movements, the transport division, and there I was confronted by a stoofy[1] flight sergeant who, when I asked about a flight to Kisumu, said, 'No way, the aircraft is full.'

I was not feeling strong from all the travelling and now just wanted to get home, so I pleaded, 'Please, I've got to go today!'

He said, 'No way.'

My tolerance was wearing thin and I demanded to see the commanding officer. When I presented the letter to the CO he read it and said to the flight sergeant, 'Lieutenant Colonel Strever is on the aircraft today. I don't care who you kick off, he goes to Kisumu today!' When I arrived in Kisumu, I bumped

THE SOFT TOUCH OF HOME

into Taffy Rees, who had four years before been my flying instructor at Induna in Bulawayo. We had a chat and then I caught the first aircraft out and eventually landed at Bulawayo feeling weak and worn out. Although weary, with excited anticipation I immediately caught a lift into town to see Bee. I hadn't seen her for three years and three months, and was about to pitch up, a week before Christmas, from out of the blue.

On the way into town, who should I pass driving down the road with an Air Force chap next to her, but Bee. As I caught sight of her, she also caught sight of me. A little uneasily I thought … oh! Oh! … but continued on to her house. I knocked at the door and Vi, Bee's sister, opened it. She looked at me in astonishment and said, 'I don't believe it!'

At this point, having left her dance partner confused and in the lurch, Bee swept in excitedly and threw herself into Dad's arms.

While Dad was doing his pilot training in Bulawayo, Bee, my mother, had volunteered as a Red Cross nurse and done her nursing training. During his long absence she had nursed hundreds of wounded and dying airmen who had been sent out to the Rhodesian military hospital in Bulawayo. Her nursing had brought the reality of the distant war horribly close to home and was a daily reminder of my father's uncertain fate. The memory of some of the seriously wounded airmen whom she had particularly nursed was burned forever in her heart and mind. She would recount them, many years later, to us, her grown-up children, with an emotion and clarity undiminished by the passing of time.

When their long embrace was over, she stood back to look at him. She later described the heart-rending shock she felt: 'He was emaciated, weather-beaten, worn out, war-torn and badly scarred. He was almost unrecognizable, and I thought to myself, is this really the same man I'm going to marry?'

However, her reaction was fleeting because, as she said, his character and sense of humour were as lively as ever. Time was short. Soon, in an excited rush and scurry, wedding plans and travel plans were in process. The marriage was to take place in Johannesburg in early January, only a couple of weeks away. Dad caught the train home to Potchefstroom, and my mother and her family followed to Johannesburg, where both families met prior to the wedding.

Four days before the wedding Dad went to the doctor to have his burn wounds checked and re-dressed. The doctor took one look at him and requested a urine specimen from him, which he thought was a bit odd, as it

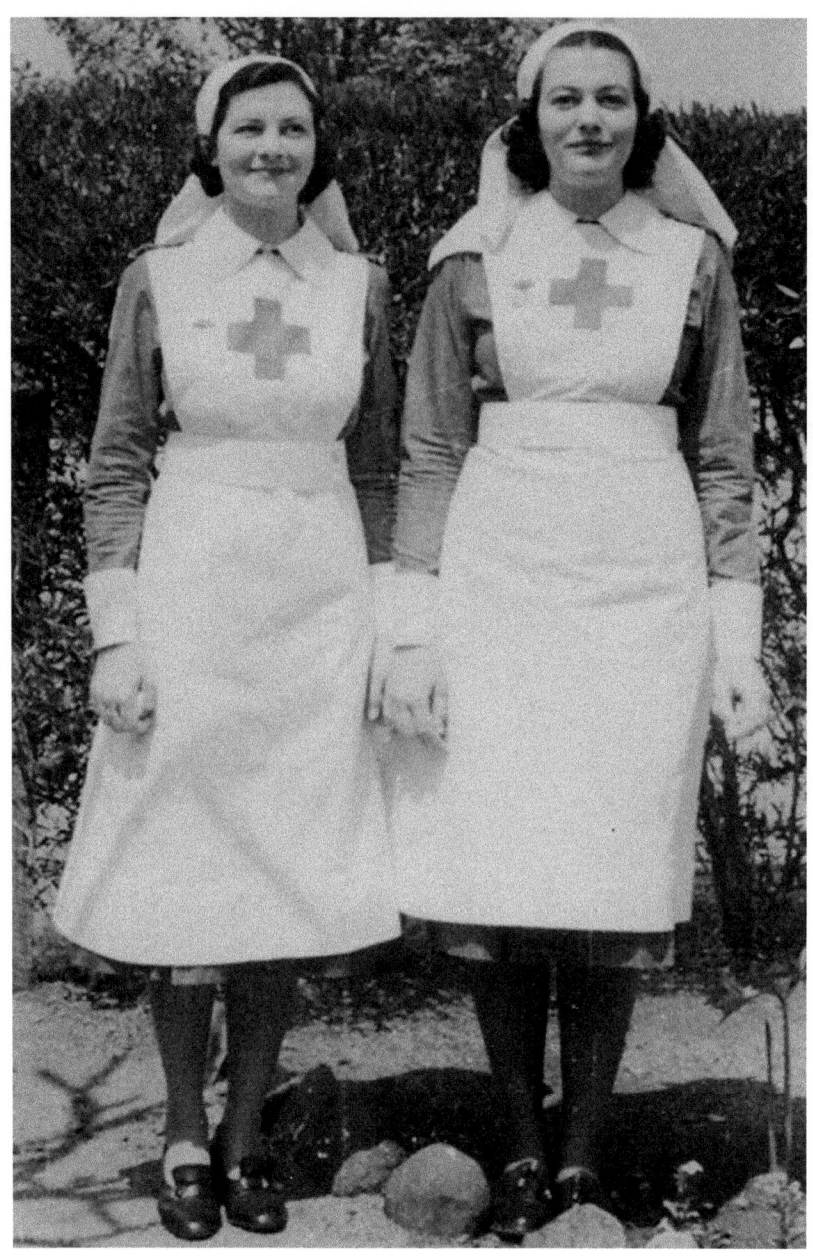

Red Cross duty in Southern Rhodesia, 1942-45. Vi, Bee's sister, on the left, and Bee on the right. They both served at the military hospital set up at the army drill hall in Bulawayo, mainly caring for injured airmen. Author's collection

THE SOFT TOUCH OF HOME

Ted and Bee swimming together before their wedding, 1945. Author's collection

had nothing to do with his dressings. Dad says: 'I brought in the specimen. It was bright orange.'

On the sight of it the doctor stated very firmly, 'My boy, you are going straight into hospital.'

Dad, horrified, replied, 'Rubbish! I'm not going into hospital. I'm on convalescent leave and am getting married in four days' time.'

The doctor insisted, 'That's too bad, my boy, you have a severe attack of yellow jaundice and are going straight into hospital.'

So into hospital he went. Once in hospital he continued to prevail on them to allow him out for a day to get married, which they eventually agreed to do. He was permitted to leave at 9.00 am on his wedding day, Saturday, 9 January 1945, and had to be back in hospital at 7.00 pm.

While he was in hospital the wedding plans and preparations were going on with some serious meddling and intervention from Dad's large contingent of aunts. His Auntie Katie insisted that Bee wear her wedding veil, a lavishly long, fine Brussels lace train with an ostrich feather headpiece, which was quite beautiful. Aunt Katie also took it upon herself to change the venue for the reception. Dad recalls being 'bulldozed into a ghastly bare little place between Troye and Nugget Street, Johannesburg.'

All the women involved were single-minded about one more detail: Dad's Air Force cap. He says, laughing, 'Oh, *ja*! What a palaver about my tatty, old dirty cap, which I had brought back from the Far East. "You can't possibly wear that!" they all insisted. "You have to go and find a new one." Well, my hospitalization made this damned near impossible, but to appease them, on my release from hospital at nine o'clock for the wedding, which was at two-thirty, here I was, going from shop to shop in Johannesburg to find a bloody new cap. I searched in vain and by ten past two I decided, to hell with this, and went ahead and wore the old one. I thought I would get flak from the women but, in fact, they seemed to overlook it!'

All went well, and the ceremony was held at All Saints Chapel of St Mary's Cathedral and performed by the Reverend Father Hewitt. My mother had lost both her brother and father to cancer in the preceding years, and so was given away by her uncle, William Rogers. She wore a wedding gown of white embossed organdie designed with a long train and, of course, Aunt Katie's wedding veil. At the reception my jaundiced dad ate poached eggs and drank ginger ale, which he describes as 'quite a peculiar wedding feast'. He was back in hospital that night, having finally wed his 'Ava Gardner lookalike' and the girl whose hair, he always said was 'her crowning glory'.

Ten days later, he was released from hospital and the newlyweds had a few days' honeymoon in Johannesburg before his return to duty in Burma. It was nearing the end of January 1945, and his hospitalization had delayed his expected date of return to duty in the Far East. This perturbed him greatly. Even though the doctor had written him a letter confirming his illness, he thought, 'It looks as if I have cooked this one up.'

After a sad farewell to his family and my mother, he boarded a flight back to Burma by the same route he had come home. For the first time in the long three and a half years that had passed he admits to feeling 'in the depth of despair, alone and homesick' when he arrived in Nairobi. The stiff upper lip of wartime survival had been fussed, kissed and caressed away. The taste of home still lingered.

By the time we got to Mombasa I was ready to throw myself back into the fray. The long trip back was relatively uneventful, except for the fact that on the leg from the Seychelles to Addu Atoll, the crew couldn't find Addu Atoll. Everyone aboard was starting to get a bit worried after twelve hours of flying around looking for the bloody place. Thank goodness they eventually found it,

THE SOFT TOUCH OF HOME

Ted before returning to Burma, 1945.
Author's collection

Ted and Bee emerging from St Mary's Cathedral, Johannesburg, after their wedding ceremony, January 1945. Author's collection

Farewell to the family before returning to duty in Burma, Johannesburg Station, 1945. From left: Bee, Ted, Esther, Pop and Ted's sister, Sheila.
Author's collection

because it wouldn't have been funny sitting around in the middle of the Indian Ocean in a flying boat. On arrival at Colombo I reported to Group, thinking I was going to get a rocket for returning late. Dear old CO Dusty Derstan, who had given me permission to go home, just said, 'Well, Ted, I know you have had a tough time and there's no problem; we will see what we can do to get you another squadron.' So I hung around for about three weeks in Command Flight, flying all sorts of different aeroplanes, but eventually got the hell in and thought, 'No, this is enough.' I went back to the CO and said, 'Look, just post me to Burma; all the other boys have gone. Never mind about a squadron for me, I'll go up as a flight commander.'

'Oh,' he said, 'will you?'

I said, '*Ja!*' Hence I got demoted from major to captain, and was posted to 211 Squadron, the one I had originally been given command of before the accident in Ceylon, this time as a flight commander.

Chapter 20

Burma: 'With Speed to the Stars'?

> *With so much experience*
> *You must surely have understood by*
> *then what Ithaca means.*
> C.P. Cavafy, from the poem *Ithaca*

I took off from Ratmalana in Ceylon, flying a Beaufighter, on 14 April 1945, carrying Warrant Officer Foxall as a passenger. I flew via Allahabad en route to 211 Squadron at Chiringa Airbase in the Arakan Province, a western coastal region of Burma near Chittagong, in what was then Bengal, now Bangladesh. At Allahabad I dropped Warrant Officer Foxall off and was given a Harvard to fly to Chiringa. On arrival on 17 April at Chiringa I was ready to take on my duty as flight commander in 211 Squadron. However, unbeknownst to me, a couple of days before my arrival, 27 Squadron had lost their CO. He had hit a vulture when coming into the circuit to land, and had died in the resulting air crash. It was due to this tragedy that I was unexpectedly given command of 27 Squadron,[1] a Beaufighter squadron based at Chiringa, and I was promoted to lieutenant colonel.

The insignia of 27 Squadron. Author's collection by courtesy of David Appleby

Well, by this time the tides were turning in the war in Burma. However, Rangoon, the main seaport and capital, was still occupied by the Japanese.

> Early in 1945, the 14th Army [commanded by Field Marshall Sir William Slim, and comprising largely of Indian and African troops] continued to advance, no longer in the jungle but in the open plains of upper Burma. Mandalay fell in March, and Slim conducted a brilliant crossing of the mighty Irrawaddy before

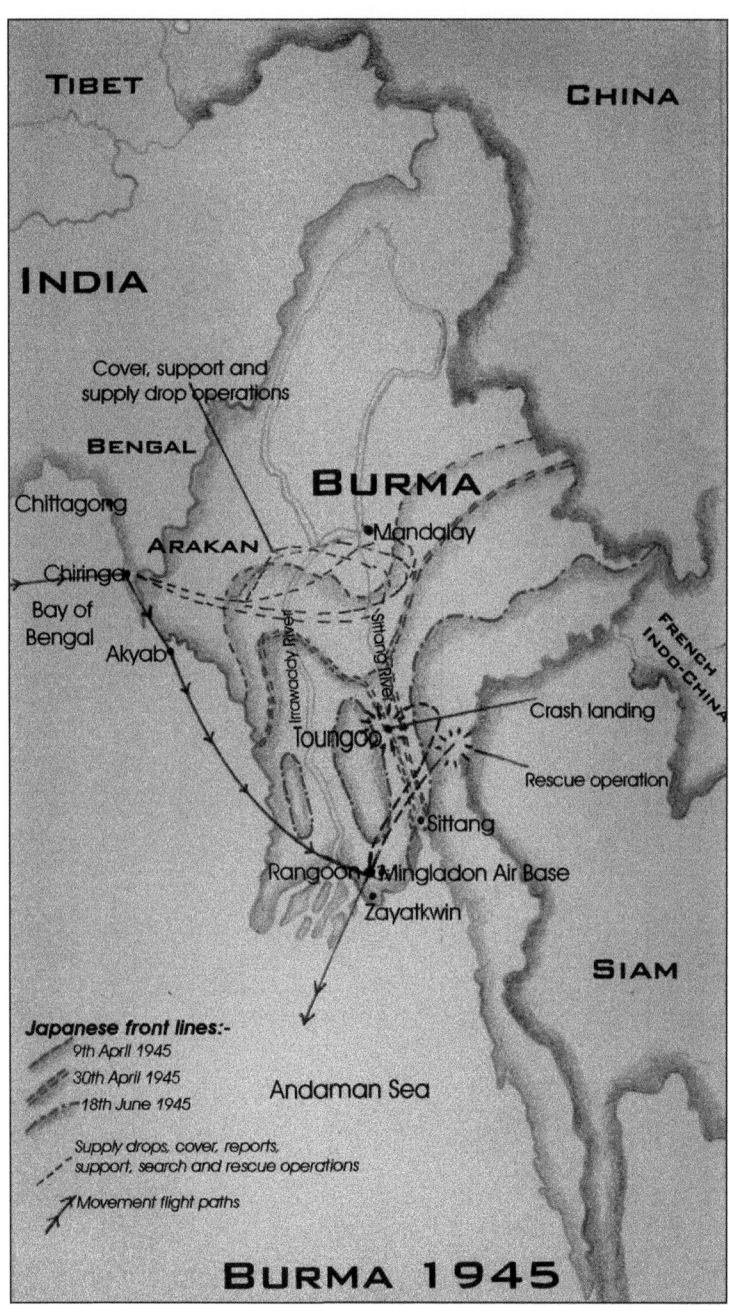

BURMA: 'WITH SPEED TO THE STARS'?

heading south. In the Arakan, the Japanese had to be winkled out of strong positions before Rangoon was taken on 3 May. Mountbatten gratified his ambition by staging a victory parade, at which he took the salute in Rangoon on 15 June. This took place despite the fact that thousands of Japanese were still fighting hard, many of them still in strength, behind British lines – as they tried desperately to escape across the Sittang River into Siam (now Thailand), losing heavily as they went.[2]

BBC History website article, 'The Burma Campaign 1941-1945'

This, then, is the arena into which Dad landed in April 1945. Before he took command of the squadron it had been employed as an anti-shipping strike wing, flying the rocket-firing Beaufighter. These operations became redundant as 1945 progressed, and when Dad took over command its operations had been changed over to ground attack, search and air-jungle rescue, still primarily flying with Beaufighters.

Dad explains ...

Now, most of the Burma countryside is like a closely-knit patchwork quilt of dense jungle, mangrove swamps, sodden rice paddy fields and mountain ranges. These patches flanked, on both sides, the mighty Irrawaddy River, which swept down from the Chinese border inland and ran straight down the centre of the country for its full length, where it eventually opened out into the sea at Rangoon. This terrain, and in addition the often lousy weather conditions, with the monsoon season lasting from May to October, was not bloody suitable to get a Beaufighter in and out of enemy territory quickly and easily. So, on taking command, I thought, 'Well, if we are going to do air rescue and supply drops, we have got to do the job properly. I want the right equipment and I don't want any uphill about it.'

As I have said, the Beaufighter was not ideal for supply drops. It was fine for searching, as it had good visibility, was good for dropping torpedoes and firing cannons. It was hopeless for supply drops because the hatch was directly below the aircraft in a difficult position to get at, which made it impossible to drop anything out of it accurately.

I had a good look at the plane. The Beaufighter had a very narrow fuselage and we sat in tandem, one behind the other. There were two bars that ran along the top inside of the fuselage, which enabled the pilot to lift himself up and over into his seat. I worked out that if we could fit bomb-release racks to the bars, and hook the supply canister to them, we could drop them through the hatch by pressing the torpedo button. However, the hatch, once opened,

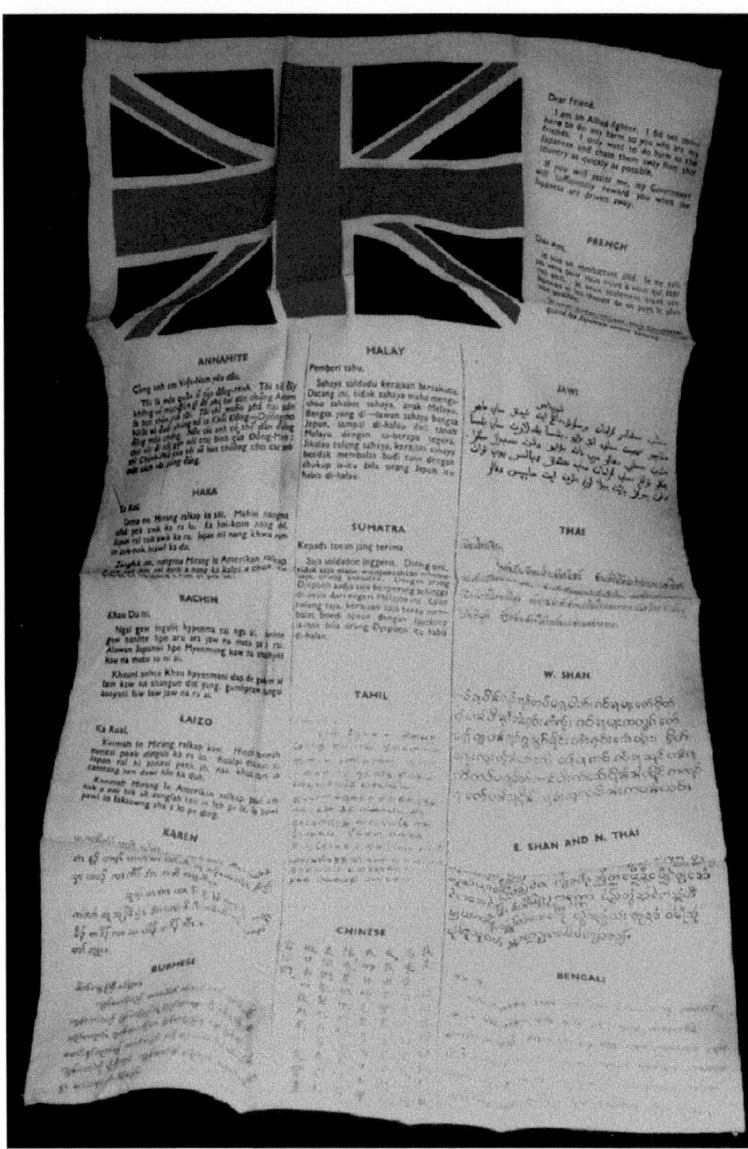

Ted's silk letter, written in seventeen languages and given to British military personnel on operations in South East Asia. It reads: 'Dear Friend, I am an Allied fighter. I did not come here to do any harm to you who are my friends. I only want to do harm to the Japanese and chase them away from this country as quickly as possible. If you assist me, my government will sufficiently reward you when the Japanese are driven away.' Author's collection

BURMA: 'WITH SPEED TO THE STARS'?

would lock, as it was in the force of the aircraft's slipstream. We would thus also need to fit a stop on the hatch to prevent it from locking open after we had released the supply canisters. With these ideas in mind I got hold of the engineer officer and described what I thought we needed. Well, they worked at this and we had a few trial and error runs and eventually it worked absolutely bloody wonderfully. Now we could do the run, hit the torpedo button, release the survival kits on target and pull out with no problems.

The other issue to deal with was to check the survival packs. I found that they had no cans of pure drinking water, as there were none available in Burma. So I instructed the stores officer to find out where we could get them. He spent a few days signalling all over the place, and found that there was a supply of them in Madras. Well, Madras was right across the Bay of Bengal, so I promptly dispatched a Beaufighter to go over and collect a large quantity of them. We then made up a quantity of packs for both sea and jungle survival.

The Beaufighter was now suitably redesigned for the supply drops and was perfect for search operations. I knew, though, that it would be hopeless for rescue operations. It was far too big, heavy and cumbersome to fly in and out of paddy fields and sandbanks. So I went down to 224 Group to see the senior air staff officer there, a chap by the name of Group Captain David, who was an old friend of mine, and I said to him, 'Look, if you want me to do this job properly I want some L5s.' The Sentinel L5 was a light and small two-seater, single-engine, tail-wheel aeroplane with a 185-horsepower motor. The seats were also in tandem, with the co-pilot seated behind the first pilot. However, the back seat could fold down, and the second control column could be removed, and this allowed for enough room to fit a stretcher in alongside the pilot, and seating room for one more person. I was soon delighted to be issued with two of them, KJ 437 and KJ 442. These two (both Kilo Juliet) I nicknamed the *Juliet Twins*. I also had requested that an expert be sent over to give us some gen on them, so that we knew how to fly them to full advantage in all situations. Of course, we had to fit long-range fuel and oil tanks to them, which we took from the Beaufighters. These we fitted on the wing struts and this gave us a fairly good range with them. These little L5s were absolutely remarkable aeroplanes; the things we did with them, and the places we flew in and out of with them! We could land and take off on river sandbanks and mushy rice paddy fields – they were fantastic! I made sure we did a great deal of practice flights with them so that we were on the ball when we were called upon for a rescue operation.

In the meantime we were already, before the end of April, engaged with supply drops to V Force in the jungle with the Beaufighters.

I had Beaufighter detachments to fly over the army units, which were advancing through the centre of Burma. Their bird's-eye advantage and presence was necessary for cover and support, and immediate reporting should a rescue operation be necessary. On 3 May our boys had taken Rangoon and reoccupied the city, as well as Mingladon, the airbase at Rangoon. At the end of May I was requested to send a detachment down to Mingladon.

On 1 June, I was contacted by someone at the headquarters of the resistance fighters of Force 136 and V Force, which was based in Ceylon. He asked, 'Can you pull three of our chaps out from behind the lines east of Sittang?' These were Indian resistance fighters of Force 136. I was a little concerned about the weather conditions, because the monsoon had started in May, and it doesn't bloody let up for months. Anyway, I replied, '*Ja*, I can.' On 2 June I took off in the rain, in a Beaufighter with Major Warren and Flight Sergeant Ganter to do an air reconnaissance of the area east of Sittang, which was 30 miles behind the Japanese lines. I flew south-east across the great Irrawaddy, over the Sittang River to the narrow area between the Sittang River and the border of what was then Siam, now Thailand. I checked all the landmarks and eventually spotted the paddy field marked out for the landing. The field was indicated by a roughly assembled bamboo X, which had been laid out on the ground by the Force 136 agents, who were waiting for us. Once satisfied that I was clear on the territory and area, I returned to base.

At this time one of the L5s (KJ 442) had gone down with fuel trouble on the way down to Mingladon, so I only had one L5 to use for this operation. As there were three chaps to rescue, and I could only carry two at a time, I decided to fly in first and pick up the first two, return to base, get refuelled and have one of my flight commanders, Squadron Leader Cameron, take over and do the second trip. Well, at least this was the plan.

The following morning, 3 June, I took off alone in the little L5 Juliet 437, at six-thirty, again in the rain. I was escorted from a distance by two Beaufighters, who were, should we be attacked, to belt everything in sight with their cannons. This they would do for both trips.

After flying for about an hour, I spotted the paddy field strip marked with the bamboo X. I descended and landed on this lot, well aware that every run made on it would increasingly churn up the marshy ground. I turned the plane around and taxied back a little way before stopping where I was sure I had enough runway to take off again. As I stopped, six Indian blokes hurtled out of the jungle and tore across to the aircraft. Looking at the size of the aircraft with consternation, one of them, waving six fingers in the air, cried out with urgency, 'We are six to go!'

I quickly replied, 'I was told only three,' also gesticulating with three fingers, and speaking as simply as I could. Their distressed looks prompted me to allay their fears with, 'No worry, I come back for others,' doing my best to gesture with my hand that I would return for them all. 'Now take two,' I said,

BURMA: 'WITH SPEED TO THE STARS'?

holding up two fingers, and got two of them to pile in on top of one another behind me. Thank God I was not the size I am today, and only weighed about 160 pounds, because with the extra weight of these two I could feel the plane sink more into the slushy ground. I started her up and with a slippery run we managed to take off all right and got back to Rangoon without incident, at eight-thirty.

As planned, Squadron Leader Cameron was ready to take over immediately for the second trip, and he took off half an hour later. I had a growing concern that, with us buzzing in and out of this place, it was becoming more dangerous. The Force 136 chaps had now told me that there were Japanese in a village only 4 miles from the rescue site, and I was sure that by now they must have become aware of us. So I decided not to send anyone else in and that I would do the third trip myself. I then waited a little anxiously for Cameron's return. Thankfully, he arrived on time at eleven o'clock with two more chaps.

After a twenty-five-minute refuelling stop, I went out on the last trip, taking off at 11.25 am. As I landed where the last two were waiting for me, without stopping, they both piled into the plane insisting, 'No time to turn, Japanese come.' Sure enough, they had become aware of us and were on their way. This meant I had a short run on which to take off and, as it happened, these two chaps were the heaviest of the group, which didn't help matters. I started my run with the wheels feeling heavy in the mud. As I was heading closer and closer to the wall of jungle at the end of the paddy field, I'm thinking, 'Come on, baby, come on baby, up you go.' Just in time, I managed to pull her off the ground and clear the thick vegetation and high trees in front of me. We returned safely to Rangoon, where the happy Force 136 chaps were keen for photographs to be taken of them posing with Cameron and me next to the little L5 KJ 437.

All the while, except for a couple of supply drops that had to be aborted because of the lousy weather, my Beaufighters were having great success at carrying out supply drops and search operations. The monsoons were now relentless and, in fact, by July, almost the whole country was flooded. However, shortly after the rescue operation in June, at a time when the heavens had really started to open, Squadron Leader Cameron came to me with some news. He exclaims, 'Oh ... Force 136 want some of their chaps dropped in again!'

I smartly replied, 'Well, this time they can drop them in by parachute. There's no way I am sending our planes back in there for any other reason but rescue operations. With these rains the bloody place will be a soggy quagmire.'

'Ah, well,' he asks, 'if I can borrow an L5 can I do the trip?'

Thinking him to be a bloody fool, but admiring his determination, I

Ted and an Indian V Force resistance fighter standing in front of an LG5 after a successful rescue operation, Mingladon, Burma, 1945. Author's collection

retorted, 'You're mad, man, but if you want to do it, you'll have to borrow an L5 from somewhere else because you are not taking one of ours.' This he managed to do and, sure enough, with two chaps sitting on one seat, he landed in soft ground and the plane promptly went straight up onto its nose, buggering up the propeller. So much for 'With speed to the stars'. And there

BURMA: 'WITH SPEED TO THE STARS'?

he was, stranded. I had to write to his family to inform them he was missing but was presumed to be OK. I explained that he was with two resistance fighters who were experts on survival in both enemy occupied territory and the jungle. Sure enough, a month later, he managed to come out unscathed.

At this time, towards the end of June, I went down with jaundice again, and felt bloody awful. This laid me up for a few weeks. Then I was given ten days' leave, which I spent at my 'home from home' with the Marches in Colombo. They were again wonderful to me. I met up with old friends, one of whom was Hugh Sheldon, who was the CO of 209 Squadron, which was flying the large Sunderland seaplanes at Koggala. Knowing I needed to get back to Burma, he offered me a lift back in one of his Sunderland aircraft, which was due to fly to Rangoon in the following two days. I was delighted to have the chance to experience the Sunderland, and jumped at it. Shortly after take-off from Koggala in this lovely big, slow seaplane, the automatic pilot packed up. As the trip was going to be a long one across the Bay of Bengal to Rangoon, I couldn't resist asking the pilot, 'Can I fly it for a while?'

'Sure,' he replied. 'You can fly it for as long as you like.'

Well, I polled this thing for about three and a half hours across the Bay of Bengal, and I absolutely loved every minute of it. When making a turn, one had to anticipate like mad. I would turn the stick, wait for about fifteen seconds, and only then it would start its turn. It was fantastic!

Back in Burma, my squadron was still engaged in many search and rescue operations and supply drops to Force 136 and V Force.

In August 1945, the last month of the squadron's operational duties, Dad's logbook records his flights as follows:

Aug 5	Spitfire VIII	Local.
Aug 5	Sentinel	Search Lib. North Akyab. Old prang.
Aug 6	Beaufighter	Search for Beau. West Mingladon. Landed Mingladon.
Aug 8	Beaufighter	Supply drop. V Force. Abortive weather.
Aug 9	Beaufighter	Supply drop V Force – Port. Eng. [port engine] Failed. Toungoo crashed.

In Dad's words ...

I took off from Rangoon. It was full cloud, with 'tents', as we called them in those days. When I got above the clouds the port engine packed up. There I was sitting uncomfortably above the cloud, no ground in sight; I couldn't go back to Mingladon because I had no radio; I had to see where I was letting down and I knew there were hills all around the place. With one engine, I needed to lighten the load; I had to dump fuel and the pack that we had underneath us.

I just managed to keep the thing on one engine just above the clouds, at about 3,000 feet. I went north, hoping the cloud would break. Anyway, I must have flown like this for about an hour and a half when I suddenly saw a small break in the cloud to my starboard side. I eased over to it, as one has to be very careful on one engine, and blow me down if there wasn't a railway line and a little aerodrome right there. So with great relief I enthusiastically shoved the nose down. Alas, in the zeal to get her on the ground I was coming in too fast. As the wheels touched down I realized that we were going to go right over the end of the runway. I pulled the wheels up so we would belly slide over, instead of nosedive into the paddy field at the end. There was an awful crunching and grinding of metal as we careered forward, eventually coming to a halt beyond the runway in the paddy field. I was very unpopular there because, unbeknownst to me, they had laid metal strips down over the soft wet ground between the runway and paddy field. My belly-landing had rather inconveniently ripped them up badly, and they now had to be repaired. This was what I did on the day the Japanese officially surrendered. However, the following day, we flew back to Mingladon in another Beaufighter.

Aug 5	Spitfire VIII	Local.
Aug 10	Beaufighter	Toungoo to Mingladon.
Aug 11	Beaufighter	Mingladon to Akyab.
Aug 14	Sentinel	Search for missing boat around Baronga. Located survivors at lighthouse.
Aug 22	Beaufighter	Air Test.
Aug 23	Beaufighter	to Hathazari – Duff weather – Returned – collapsed on landing. Tail OLED [On landing tale hit the runway and broke away].
Aug 27	Beaufighter	to Mingladon.
Aug 28	Beaufighter	to Akyab.
Aug 29	Beaufighter	To Mingladon.
Aug 30	Beaufighter	Supply drop – V Force. Successful.
Aug 31	Harvard	Zayatkwin to Hmawbi.

BURMA: 'WITH SPEED TO THE STARS'?

The last trip in August to Zayatkwin, a Spitfire airbase about 40 miles south of Rangoon, was made on a request that came through from a Spitfire squadron based there. Dad's squadron had been asked to provide three Beaufighters as an escort to a squadron of Spitfires, which was to fly to Singapore. A navigational Beaufighter would lead the squadron; two Beaufighters would fly as observers in the event of any problems, on either side of the formation to the rear. If a Spit went down one of the Beaufighters was detailed to drop in a survival canister and pinpoint the location of the accident for rescue purposes. This caution was felt necessary because the Spitfires had long-range tanks that on occasion got airlocks, causing them to cut off the fuel supply.

Dad says ...

Well, now, into September we shot off to this little Spitfire strip and sat there for about eight days. They would not let the Spits fly unless the weather was perfect, and this was frustrating because I felt on numerous occasions that we could have got through to Singapore. At Zayatkwin, our private war was fighting off multitudes of the largest and most ferocious mosquitoes I had ever experienced. They came boring in like Kamikaze pilots. We had to dress under our mosquito nets, having to get up at half past two every morning for briefing on the weather, as the plan, weather permitting, was to take off at first light. Anyway, while we waited there I was still doing a fair amount of flights to-ing and fro-ing between Mingladon to Zayatkwin to check on things in my own squadron. It was on one of these trips that a signal came through for me to report to Headquarters in Ceylon pending reshipment back to South Africa. So that was the end of my operational flying.

On 5 October I boarded a pocket aircraft carrier, the HMS *Khedive*, at Ceylon, and sailed home across the Indian Ocean. The atmosphere was full of great joy, but at the same time heart-rending sadness for those who would never return. It seemed unfair that I was going home when so many friends had not come through it. I felt especially upset about the loss of two people to whom I felt deeply indebted for saving my life. The South African pilot, Frank Robertson, who had flown Boyd Gerny, Corporal Edwards and me to hospital in Colombo after our terrible crash at Vavuniya, had – exactly a month and two days later – himself been badly burnt in an air crash and was also brought to the 35th General Hospital, where I was still recovering at the time, and where he very sadly died a month later. Penny Clarke, my most devoted nurse who pulled me from death's door in hospital with her loving and committed care, had also died in an air crash on her way home to England.

Aboard HMS Khedive, *homeward-bound for South Africa. Ted second row from front, ninth from right.* Author's collection

The burden of survival lay heavy on me during the trip home. There seemed to me to be such divine injustice in their loss. Questioning thoughts ran through my mind. What was it that got *me* through such narrow escapes with death? Why was *I* so blessed? I must have had a guardian angel watching over me. Or was I just bloody lucky?

As Herschell Reilley wrote to me about pilots who survived in spite of the odds: 'They had all come to the same conclusion – that there must have been an extra crew member on board … some divine assistance.' I know that the weight of this grief and questioning gave way in Dad to a feeling of great responsibility to do something meaningful in his life.[3]

BURMA: 'WITH SPEED TO THE STARS'?

The ship touched home soil at Durban on 17 October 1945 to the welcome of a jubilant crowd. I was swept up in the busy and frenetic activity and chaos with everyone trying to get home. After disembarking, I caught the night mail train from Durban to Johannesburg. It was the most wonderful feeling to be back on South African soil again and that finally, after six years, it was all over. The station was buzzing and crowded with people and, as I ran down the platform looking for a seat, I heard a voice shout out of one of the windows, 'Oh *Got*, oh, Strever.'

I looked around and said, 'Oh *Got*, oh, Lever.'

Jack Lever and I had trained together in Bulawayo, we had gone across to England together, and he had been in Hugh Sheldon's Sunderland squadron in Ceylon. Our greeting call was a hangover from our training days, when we were mere boys and pranksters.

Oh boy, did we let ourselves go and have a binge on that train. Someone in the carriage had a supply of whisky. As wartime trains don't have cups, water or glasses, this whisky was continuously passed around and slugged neat from the bottle until we were all absolutely motherless. I arrived, seriously hung-over, in Johannesburg the next morning, but my hangover didn't interfere with my urgent need to get home. As quickly as possible, I got on a train to Pretoria to report into SAAF Headquarters, and after reporting in I asked for immediate leave to get back to Bee in Bulawayo for what I knew through correspondence was the imminent birth of our first baby. I was granted the leave, but as trains were so booked up at the time, it frustratingly took me two days to get a booking.

I eventually arrived in Bulawayo on 21 October to an excited welcome from Bee and baby still intact and her family, with whom she was living. The next day, with no sign of baby budging from her cocoon, we spent the only day we would have alone for a very long time. On my second day back home, 23 October, we became the proud parents of a bonny baby girl, who we named Carolyn Mary. Her timely arrival, at Sister Roux's Maternity Home, had me secretly convinced she had waited for her dad to come home.

On 9 April 1947, Dad stood in a line on the lawn of Government House, Salisbury (now Harare), waiting to be presented with his DFC by King George VI. Pinning the medal onto his chest, the king simply asked, 'Where did you get it?'

Dad replied, 'Malta, Sir!'

'Jolly good show,' was the king's reply.

Dad often joked with us, saying, 'Ah, well, you know the whole royal family was there on a state visit to Southern Rhodesia. Princess Elizabeth was only twenty-one and Princess Margaret was younger and was a really beautiful girl. After the investiture, the royal family mingled for a while with us at the tea on the lawn. A week or so later, the royal family visited Bulawayo. I took part in the ex-servicemen's march-past parade, which of course we did in full uniform, looking very smart. As we marched past the royals my eye momentarily caught that of Princess Elizabeth, and I'm telling you, she gave me the glad eye!' Did we ever tease him about his wishful thinking!

The peacetime military was, as Dad firmly stated, 'Definitely not for me!' So when he was offered the opportunity to hold a permanent position in the South African Air Force, he declined. His passion for flying, however, didn't diminish. Until he could afford his own light aircraft, and in order to be in the air, he became a flying instructor at Induna for a few years after

BURMA: 'WITH SPEED TO THE STARS'?

Ted and Bee with baby Carolyn at a picnic at Matopos, Bulawayo, 1946. Author's collection

King George VI and Queen Consort with Princess Elizabeth behind them, arriving at Government House, Salisbury, for the investiture ceremony, 1947. Author's collection

Ted and Bee at Government House, Salisbury, for his investiture, 1947. Author's collection

the war. Here, after work and at weekends, he put pupil pilots through circuits and bumps, stalls and spins, steep turns, and cross-country flights. One day he proudly took two-year-old Carolyn with him to the airfield. However, he says, laughing, 'When I wheeled the pram towards the plane she took one look at it and screamed her head off! She just didn't want to know this strange looking thing.'

It took nineteen years before he was able to buy his own aeroplane. Of course, his story doesn't end here. He continued to live a varied, adventurous and active life in the only way he knew how: always full-bodied and warm-blooded, quite often hot-tempered, but mostly big-humoured.

Ted's log book. Author's collection

Epilogue

> All love lives by slowly moving towards its end,
> and is sharpened by the snake-bite of farewell within it.
> Laurie Lee, from *Two Women*

A five-year-old boy, travelling through the Transvaal Bushveld, perched high on top of the folded canopy of *Molly*, and dreaming of flying, had more than lived out his dream. Dad always professed that flying gave one a different perspective on life and the world.

In *A Gift of Wings*, Richard Bach, writing about the reasons pilots love to fly states, 'One reason is the finding of life itself, and the living of it in the present.' He goes on, 'The terms that flying lays down for pilots [are]: Love me and know me and you shall be blessed with great joy; Love me not, know me not, and you are asking for real trouble.'

This book is the first part of three phases of Dad's life. The second part is his life in Southern Rhodesia, where, in 1945, the promise of peace and stability would last for a few years, and where he spent the next twenty-five years building up a business and raising four children.

The third part has him back in South Africa, where he had the opportunity to settle and live in Magoebaskloof, the place that had caught his imagination as a child and where he spent the remaining twenty-seven years of his life. His life had come full circle. Here, in 1976, after thirty-one years of a devoted albeit at times tumultuous marriage, Bee, my mom, died after a long illness. Dad was lost and bereft; she had been, in his words, 'my strength and mainstay'. My mother had been his passionate and nurturing companion, steady through his life's disasters, and backstage organizer for his exciting adventures. A few years later, he met and married Andrée Stanford, left his farm in the Magoebaskloof Valley where he and Mom had lived, and resettled on Stanford Farm at Lakeside, near Haenertsburg.

Around this time, at a party in Magoebaskloof, two old veterans of the Air Force, Clifford 'Potty' Thompson and Dad, were introduced and got chatting – forty years after the war. They discovered that they had both been in the SAAF and had been seconded to the RAF.

Ted asked Potty, 'Where were you stationed?'

'Malta,' was Potty's reply.

EPILOGUE

Stanford Farm, Magoebaskloof, early 1990s. From left: Ted's beloved Staffy, Socks, Ted, my daughter Abigail, myself, Andrée, and my husband, Michael. Author's collection

Dad, raising his bushy eyebrows in surprise, said, 'Good Lord! So was I. I was with 217 Squadron on torpedo bombers at Luqa Airbase.'

'Heavens! I was in 86 Squadron, also at Luqa. However, my operations from there ended when I was shot down and ended up in an Italian POW camp.'

'When were you shot down?' enquired Dad.

'On 29 July '42. We were damn lucky because the crew and I all survived the ditching. We managed to get into the dinghy and watched the plane slip under the waves. However, it wasn't encouraging when we opened our survival packs because they had little in them. [The locals on Malta had been known to steal food and water from the ration packs.] We were shot down somewhere near the Greek coast so we decided to paddle towards it. On that first day we managed, with great relief, to get close enough to the coast to see the setting sun reflecting off the windows onshore. However, we woke the next morning, the thirtieth, to find that during the night a strong current had damn well swept us way out to sea again.

'So there we sat, with minimal rations and horribly exposed to the summer sun. By nightfall we were quite perturbed so we decided that if we saw a ship, enemy or not, we would fire our flare. At about two o'clock in the morning we nearly got run down by a blacked-out ship and lost no time in firing our flares. The ship made no acknowledgment and continued on her way. I was convinced by the third day that we were not going to survive unless we were found very soon, the chances of which were very slim. We were also getting dreadfully sunburnt and blistered. On the 31st, at about eleven in the morning, we saw an Italian warship in the distance, carrying out a square search, and after a few hours she duly picked us up. I asked the crew on the ship if they had found us because of the flare we had fired the previous night. "No," said the Italian skipper, "we are searching for a Cant floatplane of ours that went missing two days ago in this area."'

'That's very interesting,' exclaimed Dad. 'You see, I had been shot down on 28 July, the day before you, and we were also picked up by an Italian floatplane, and taken as POWs. On the following day, they were flying us from Corfu to Taranto to POW camp in a Cant Z.506B and my crew and I managed to overpower our captors and took over the plane. I then flew it back to Malta. That must have been the Cant they were looking for!'

'Good God! You damned well saved my life!' declared Potty, incredulously.

Every year after their first meeting they would celebrate 28 July together with a ritual dinner and drinks. They both died in Haenertsburg, in the same year, and at the same age.

There were a few highlights in Dad's last few years. One was a trip to his old squadron, 27 Squadron, in the United Kingdom in 1992. He was accompanied to the base at Marham by friends and fellow veteran fliers Dudley Cowderoy and Norman Hearn-Phillips. They were warmly welcomed by members of the squadron, and given a tour of the base. On the tour my very large father was enticed, by two lithe young jet pilots, into the tight-fitting seat of a Hawker jet, where to their consternation he became firmly lodged. Dad says, chuckling, 'It took some strong young arms and a bit of a bloody battle to get me out of the thing.' The squadron then treated him to a surprise VIP lunch, with the whole squadron in attendance. During the speeches Dad was able to say a few words in what Dudley Cowderoy

EPILOGUE

The mess at Chivenor, with members of 27 Squadron, 1992. Dudley Cowderoy on the far left. Author's collection

describes as 'his splendid way' to all the young pilots. From this visit he came home with presentation photographs of the squadron's emblem, the statue of an elephant on a plinth, around which is inscribed all the names of the past commanders.

When shown the photograph, his wife Andrée piped up, laughing, 'Oh God, you're under the elephant's arse!'

Dad's response was a resigned, 'She leaves me no dignity.'

Another highlight, also in 1992, was the receipt of an application form by the Malta High Commissioner to the United Kingdom with the recommendation that Dad apply for the Malta George Cross Fiftieth Anniversary Medal. The instructions for the application stipulated that details given must be corroborated by official documentary evidence of service and must be accompanied by a photocopy of such evidence. Dad duly sent off details of his service to the RAF Personnel Management Centre

Ted lodged in a Hawker jet during the visit to 27 Squadron, Chivenor, 1992. Author's collection

in December 1992 requesting the required information. The astonishing response from the British Ministry of Defence was: 'It is not possible to state categorically that you served in Malta. As you can see from the enclosed Record of Service ...' Although Dad served in Malta, which is clearly documented in the official records of 217 Squadron's activities and strikes from Malta and the well recorded 'skyjack' episode, he was, of course, never officially posted to Malta. He and his crew were commandeered there en route to the Far East.

The official award ceremony took place in early 1993 at the Maltese Embassy in London, where many of his comrades-in-arms gathered. Among them was the Royal Navy captain of one of the two destroyers that heroically carried the lifesaving *Ohio* into Valetta harbour between them in what the Maltese named the 'Santa Maria Convoy'. Dad, deeply disappointed, missed the event. Indignant at his 'non-existent' status on Malta, he continued to battle it out with the RAF authorities. Finally, a year later, he received a letter from the office of the prime minister of Malta stating, 'I have the honour to inform you that the President of Malta has been pleased to approve

EPILOGUE

The elephant emblem statue of 27 Squadron with Ted's name inscribed under the elephant's tail! Author's collection

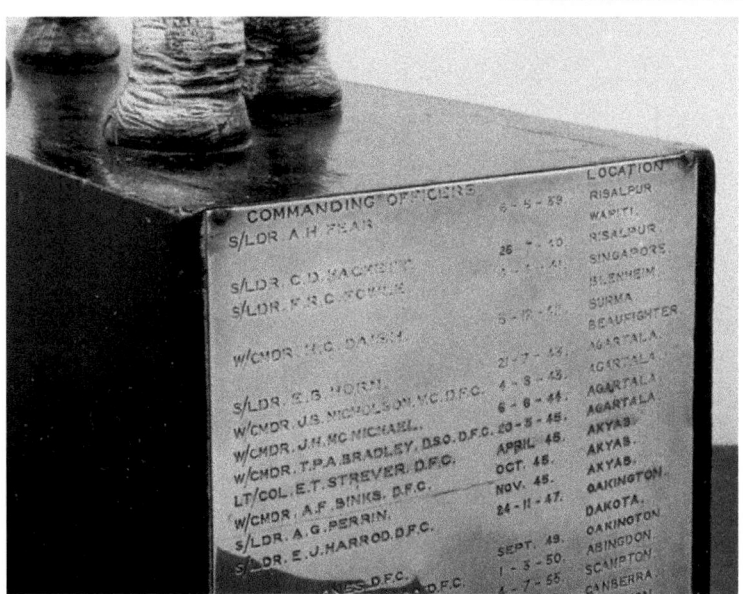

Ted's medals, showing the Malta George Cross Fiftieth Anniversary Medal, second from left.[1] Author's collection

the Prime Minister's recommendation that the "Malta George Cross Fiftieth Anniversary Medal" be awarded to you. It is, therefore, with great pleasure that I am herewith forwarding to you the "Malta George Cross Fiftieth Anniversary Medal".'

Fifty years to the month after the skyjack episode, Dad was sitting at his desk opening the usual mail when he came across a letter posted from Turin, Italy. Written in elegant old-fashioned Chancery Cursive, the letter was from a veteran Italian airman named Riccardo Alessio. All those years ago, Ricardo had witnessed the arrival of the Italian Cant when it landed with the British POWs at the Regia Aeronautica Airbase at Prevesa, Greece, on 28 July 1942. Enclosed with the letter was a photograph of the Italian crew and their prisoners arriving at the quayside of Prevesa. The letter extended warm and friendly greetings, and a polite request that Dad identify and name himself and his crew in the photograph before returning it to the sender. Through the correspondence that followed between them, Dad learned that the first pilot, Mastrodicasa, and second pilot, Chifari, had carried on with their military careers. Mastrodicasa had attained the rank of general, and both were at that time in their eighties. In a poignant and eloquent letter to me after Dad's death, Riccardo Alessio wrote about his contact with Dad:

EPILOGUE

A sense of friendship was born this way, as is natural between men who have confronted, with the most utmost spirit of sacrifice, the suffering of a terrible period of war: fighting with the highest sense of honour, but <u>never</u> with hate.

The last time I watched Dad fly, he had taken my children, Abby and Ben, for a flip over Stanford farm, which always thrilled them. It was a breathtakingly perfect day. I waved from the ground as they flew over and around the cottage. His little silver aeroplane sparkled in the sun as it waggled its wings at me and playfully flipped about in the radiant cobalt sky. Gazing upwards, a sudden, almost sacred, sense of knowing and understanding engulfed me. Up there, Dad, unbounded by the weight of earth or time *was* the little silver aeroplane; he became completely at one with it. The image of a glittering and dancing little aeroplane deeply etched itself on my soul's eye that day.

Excerpt from Riccardo Alessio's letter. Author's collection

This wasn't the last time he flew – indeed, he remained very firmly at the controls until the age of seventy-five, when he finally submitted to time and sold his aeroplane.

On 27 January 1997, our recording sessions came to an end. Dad, in a whisper, said to me, 'I can't talk anymore.' On 18 February 1997, Dad left 'the departure lounge' and took his last flight. He had survived eight air crashes and had never lost a passenger. He died at home in Stanford Cottage

Messina Airport, South Africa, on a flying trip to Zimbabwe in 1992 with Ted. From left: Ted, Abigail, Gail, Benedict and Michael. Author's collection

in the midst of much love. True to character, he had given all sorts of instructions about his funeral: no fancy black hearse, just a farm *bakkie* (to find one long enough for him proved to be quite difficult!). He insisted on a plain pine coffin with rope handles. He chose the hymns, to which his wife Andrée's response was, 'Oh God! He's chosen all the tear-jerkers!' He had called for Piet Sehkwana, his factory driver, who was also the factory choirmaster, to come to the cottage to see him. He had a soft spot for the factory choir because he had watched and supported their development and their success in coming second in the National Choir Competition. When Piet entered the bedroom and stood at the bedside, Dad said, 'Now Piet, I am not going to be here for long, so I want you to get the choir to practise so that they can sing at my funeral.' Piet gave his assurance that he would do this and, with tears in his eyes, shook Dad's hand for the last time and left the room.

The tiny Haenertsburg Catholic church, bright with colourful flowers, like a cup overflowing, spilled family and friends from its lip onto the lawn outside. After the suffocating sense of solitude we had felt, brought on by weeks of mist and rain, the day of the funeral was bright and clear.

'What a beautiful day for flying!' Father Willie's opening words said it

EPILOGUE

all. 'Today,' he continued, 'we accompany Ted on his last flight, the most important one he has ever made ... he has had to hand over the controls to God ... that is never easy for a good pilot like Ted ... [he] never lost the art of living fully from his centre ... from his spirit, which is from the creative hand of God.'

These words immediately brought to my mind one of my own, and Dad's, favourite poems, *High Flight*, by John Gillespie Magee,[2] which Dad had written out and sent to me in a letter many years ago:

> *Oh! I have slipped the surly bonds of Earth*
> *And danced the skies on laughter-silvered wings;*
> *Sunward I've climbed – and joined the tumbling mirth*
> *Of sun-split clouds – and done a hundred things*
> *You have never dreamed of – Wheeled and soared and swung*
> *High in the sunlit silence. Hov'ring there*
> *I've chased the shouting wind along, and flung my eager craft*
> *through footless halls of air ...*
>
> *Up, up the long, delirious, burning blue*
> *I've topped the wind-swept heights with easy grace*
> *Where never lark, nor ever eagle flew –*
> *And, while with silent, lifting mind I've trod*
> *The high, un-trespassed sanctity of space,*
> *Put out my hand, and touched the face of God.*

The eulogy, written and read by Frank Tilley, the son of Dad's lifelong friend, Don Tilley, started:

> I'd like to compare the life of our friend Ted Strever to a great book. It is one of those books that we would eagerly read at every opportunity. Such was the character of Ted, and the fullness of his life, that we selfishly would have wanted the book to never end.
>
> This great book – as with his life would be filled with many chapters – chapters of excitement and drama; loyalty and love; highs and lows – it would have chapters of dogged determination, and many of humour and happiness – and for me all the chapters would be bound in a cover of warmth and friendship. ... There always seemed to be fun, excitement and

> **High Flight**
>
> Oh, I have slipped the surly bonds of earth,
> And danced the skies on laughter silvered wings;
> Sunward I have climbed and joined the tumbled mirth
> Of sun split clouds – and done a hundred things
> You have not dreamed of – wheeled ~ soared ~ swung
> High in the sunlit silence.
> Hov'ring there
> I've chased the shouting wind along & flung
> My eager aircraft through footless halls of air.
> Up, up the long, delirious, burning blue
> I've topped the wind-swept heights with easy grace
> Where never lark, or even eagle flew;
> And while with silent, lifting mind I've trod
> The high untrespassed sanctity of space,
> Put out my hand and touched the face of God.
>
> JOHN GILLESPIE MAGEE (Jnr)

High Flight written out by Ted in a letter to the author, Gail, his daughter. Author's collection

happiness when Ted was around. He had a great ability to make you feel as though life was great. I am sure all of you battled to get Ted to talk about himself. He would always stress the role others had played. Yet, if ever there was a need for assistance, time or effort, Ted would be first in line to tirelessly commit himself to the cause – he gave so much and didn't expect anything in return.

EPILOGUE

Just in case you think I have forgotten those other chapters, those ones in which Ted would dig his heels in and refused to budge: you all know he was sometimes one of the most stubborn and frustrating people on earth. He took great delight in debate, but I think he often took a particular stance to fire up someone else. It was at times like these that my late mother had some rather choice language to describe him. Ted was a man of conviction – he had guts and determination. He stuck to the task and whatever he believed in. Finally, Ted, up there I am sure you will have once again earned and been awarded your wings and flying licence. As an angel I am not sure how good your singing voice will be, but I bet you'll have the happiest and loudest laugh in heaven.

Frank also related how, when he was five years old, when Dad was visiting he would lie awake in bed and wait for the 'booming and very contagious laugh to explode in the lounge and rumble down the passage to my bedroom. … It seemed that I never had bad dreams when Ted Strever came to dinner.'

After the service, we all trundled up the rocky dirt road to the cemetery on the hill above Haenertsburg. The factory choir, which Dad had always been so impressed with and proud of, were assembled around the grave and sang *Nkosi Sikelel' iAfrica*, their harmonizing voices carried by the breeze floating above us and over the village at the foot of the hill.

At his final resting place, the burial rites were performed by Father Willie. As Dad had specifically requested, my husband Michael said the final prayer at the graveside because, as Michael joked, 'He thinks I have a hotline to heaven!'

The South African Air Force Association flag, which draped the coffin, was removed, and the coffin was lowered into the grave.

We all stood with unfocused gaze directed at the hole in the ground, now slowly being filled with soil, shovel by thumping shovel. It was a rooted and holding moment, when suddenly there was a rumble in the sky. We looked up. A tiny black speck emerged from a cloud. A South African Air Force Cheetah jet spearheaded and hurtling through space did a wide circle above us and then descended. Swooping down on us it shattered the moment apart, tearing one world from another. The earth shook with its surging power, and the roofs rattled in the village below. Twice it slashed through the wall of our deaf silence, sending shivers up our spines as it skimmed the tree tops in a final salute to one of its airmen, somehow making the separation complete and dropping a curtain on Dad's life. Its final ascent dragged with it the booming sound, and left us in stunned silence.

> *'tis grace hath brought me safe thus far,*
> *And grace will lead me home.*
> John Newton, *Amazing Grace*

The laugh! Ted relaxing on Stanford Farm, Magoebaskloof, in the mid-1990s.
Author's collection

Gail with Ted in Magoebaskloof, 1984.
Author's collection

Appendix

It is not the critic who counts, not the one who points out how the strong man stumbled or how the doer of deeds might have done them better. The credit belongs to the man who is actually in the arena; whose face is marred with sweat and dust and blood; who strives valiantly; who errs and comes short again and again; who knows the great enthusiasms; the great devotions and spends himself in a worthy cause and who, if fails, at least fails while bearing greatly so that his place shall never be with those cold and timid souls who know neither victory or defeat.

President Theodore 'Teddy' Roosevelt, from 'Citizen in a Republic' speech at the Sorbonne, Paris, 1910. Miller, Nathan, *Theodore Roosevelt: A Life*

Major Frank Robertson
Major Frank Robertson died on 3 January 1945, aged thirty-six, in a flying accident off the coast of Ceylon. He is commemorated at the Liveramentu Cemetery near Colombo, Sri Lanka, commemoration number 2F.7.

Lt Col D.P. (Don) Tilley DSO and DFC with bar, SAAF, seconded to RAF 1941
Don Tilley served in the Middle East and Malta and was highly acclaimed as one of the most daring torpedo bomber pilots for his undaunted courage when pressing home attacks against the enemy. He was nicknamed 'Lucky Tilley' by his fellow fliers for surviving a record number of torpedo drops. As a member of a small elite group of airmen who became known as the 'shipbusters', he was also renowned for sinking the indestructible ship *Kuckuck*, an enemy minelayer. He was, at the time, the commanding officer of 19 Squadron. Also read *The Ship-Busters*: *The Story of the RAF Torpedo-Bombers*, by Ralph Barker.

Lt Col E.G. (Gill) Catton, DSO and DFC, SAAF, seconded to RAF 1941
Gill Catton was the commander of 19 Squadron in the Middle East and Mediterranean prior to his return to South Africa when Don Tilley took command of the squadron. Catton earned his awards for his many anti-

submarine and anti-convoy strikes in the Mediterranean. In 1944, his squadron, based in Italy, made an astonishing number of attacks on German-held targets in Albania; many strikes were led by Catton himself. His squadron's persistent offensive on German strongholds in Albania, aided by the Albanian partisans on the ground, was responsible for pre-empting the German withdrawal from Albania in late 1944. At the end of his tour of duty Catton had achieved eighty-six sorties on enemy targets

Cliff Evans
Lieutenant Clifford Evans was born in Durban in 1915, and after joining the SAAF he was seconded to the RAF and served with 39 Squadron on Malta. He was reported missing at 13h20 on 6 September 1942, when his Beaufort Mk II failed to return from an anti-shipping patrol north east of Ducato off the Greek coast. He has no known grave and is therefore remembered on the Malta Memorial, Floriana Area, Valetta, Malta. He was twenty-seven years of age at the time of his death.

Gordon Brodziac
Lieutenant Gordon Brodziac was born in Cape Town in 1915 and after joining the SAAF was seconded to the RAF, where he served with 608 Squadron on Malta. He was reported missing on 11 November 1942, when his Lockheed Hudson Mk III aircraft failed to return from an anti-shipping strike. He has no known grave and is remembered on the Malta Memorial at Valetta. He was twenty-six years of age at the time of his death. His name also appears on the Bishop's Roll of Honour (Bishop's Diocesan College, Cape Town, South Africa).

Willy Wilson
Lieutenant William Wilson was born in Boksburg in 1915. After joining the SAAF he was seconded to the RAF and served on Malta. He was reported missing on 4 July 1942, at sea off Luga, when his Beaufort Mk I was shot down into the sea while on a transit flight from Malta to the Middle East. He has no known grave and is therefore remembered on Column 267 of the El Alamein Memorial, in the El Alamein War Cemetery, Egypt.

Carey Heydenrych
Heydenrych was shot down in action and became a POW at Stalag Luft III at Sagan in Silesia. He participated in what became known as the 'Great

APPENDIX

Escape', which was masterminded by another South African pilot, Squadron Leader Roger Bushell. At the time of the escape, Heydenrych was already in the tunnel waiting to exit when shots were heard in the woods, and the escape was aborted.

Wally Morgan
Wally Morgan was shot down and taken prisoner, and was also a POW with Heydenrych in Stalag Luft III's eastern camp. He suffered from post-traumatic stress disorder throughout his life.

Bob Rogers
Bobby Rogers later became Lieutenant General R.H.D. (Bob) Rogers, Chief of the South African Air Force, and in receiving the Order of the Star of South Africa, he became the most highly decorated officer in the SADF at the time.

Raymond De Marillac
Raymond was killed in a training accident at Heaney, near Bulawayo, in 1941. Shortly before he was killed he had been badly burnt in trying to rescue a pilot from his aircraft, which had crashed on landing and burst into flames. Raymond rushed to help and, in trying to release the pilot's parachute harness, he was burnt on the face and arms. When he tried to turn the quick-release handle, it was so hot that it bent in his hand, burning his hands badly. He then had to abort his attempt to save the pilot.

Notes

Acknowledgments
1. From *A Course in Miracles* by Helen Schucman and William Thetford.

Prologue
1. Los-kop is an Afrikaans expression, which literally means 'loose-head'. The correct way to say it is kop-los (head-loose), because Afrikaans say things backwards. Los-kop is used here because English speakers in South Africa usually mistakenly put the adjective first.
2. See Appendix.

Chapter 1
1. 'Thank goodness we hadn't had our undercarriage' refers to the crash-landing, implying it would have been a lot worse with the undercarriage down.
2. Records from RAF archives.
3. **11-16 June 1942**
Operation *Harpoon*: (These convoys were heavily attacked for four days.) Two merchantmen reach Grand Harbour, Malta. Losses: one cruiser, five destroyers and four merchantmen. They delivered 25,000 tons of supplies, saving Malta from starvation for a further two to three months.
Operation *Vigorous*: convoy from Alexandria comprising eleven merchantmen, one dummy battleship, eight cruisers, twenty-six destroyers, four corvettes, two minesweepers and two rescue ships forced by presence of Italian fleet to return to Egypt. Losses: one cruiser and five destroyers; no supplies delivered.
From *Malta: The Thorn in Rommel's Side*, by Laddie Lucas.

Chapter 3
1. Grandpa Higgins swore that the Jonker Diamond was the diamond that the old chap had lost. The story of the find is as follows: Johannes Jacobus Jonker a sixty-two-year-old digger was working a claim at Elandsfontein, 4.8 kilometres south of Premier Mine, about 40 kilometres east of Pretoria.

NOTES

On 17 January 1934 he sent his son Gert along with two of his native South African employees to direct operations on the claim. One of them, Johannes Makani, was washing a bucketful of gravel when he suddenly stopped dead in his track and picked up something. Without saying anything he walked to the cleaning camp and scrubbed the object he had found, which had been caked with dirt. He then threw his hat in the air and shouted, 'Oh God, I have found it!' From .tripod.com/jonkerdiamond.html
2. The eighteenth-century statuette in Brussels of a small cherub-like boy peeing from a raised stuccoed marble plinth into a fountain below.

Chapter 4
1. Bennie Osler was a Springbok rugby fly-half in 1930.

Chapter 5
1. A 'tickey' was a small silver coin worth three pence, so half a sixpence.
2. 'Cuts' was the term used for strokes of the cane.
3. Dr Taylor-Smith was First Medical Health Officer for Pietersburg, but in 1935 went into private practice. From Louis Changuion: *Pietersburg; Die Eerste Eeu*, p.135.

Chapter 7
1. See appendix.

Chapter 8
1. See appendix.
2. See appendix.

Chapter 10
1. See appendix.

Chapter 13
1. See appendix
2. See appendix
3. Goose-neck flares: loose, paraffin-lit, flare cans with a wad of wool sticking out of a spout, which gave off a big, black, smoky flame.

Chapter 14
1. These were the distinctive red tabs worn on the South African Air Force

uniform; although seconded to the RAF, Dad continued to wear his SAAF uniform, wearing the RAF uniform only for formal or ceremonial occasions.

2. This, we realized later, was because they had not been necessary; the Italian crew were going home on leave, and the aircraft had been destined to go in for maintenance.

3. The pilot, Bill Young, was to become a good friend and also settled in Bulawayo after the war.

Chapter 15
1. The Beaufort Mark I, which Ted had been flying, had British Bristol Taurus engines, but had fixed propeller blades, so the pitch of the blades could not be 'feathered' (turned edge-on to the wind flow). This meant that in the case of engine failure the propeller blades caused an enormous amount of drag. The Beaufort Mark II had American Pratt and Whitney variable pitch engines. This meant that the propellers could be swivelled to create the least wind resistance. It made for better cruising, added 25mph of speed and improved the odds in emergencies.

Chapter 17
1. Herschell Reilley describes how unlikely it is that a main wheel would become detached from the plane when the undercarriage is up: 'When flying, it is tucked in behind the engine. Then, two flaps cover it, and it is firmly fastened by the oleo legs, which support the plane. The whole position of the wheel should never cause it to break away, as the engine would take the shock of ditching. My suggestion is that they had some divine assistance. I have been with a group of fliers who had all been on ops ... and they all knew of unexplainable times when those who should have perished did not do so. They had all come to the same conclusion – that they must have had an extra crew member on a certain flight.'

Chapter 19
1. Ted often used the word 'stoofy' to describe someone he thought arrogant or above themselves with a difficult attitude.

Chapter 20
1. 27 Squadron's emblem is the Indian elephant crowned and encircled with the motto, *Quem Celerrime Ad Astra*, 'With Speed to the Stars'. The

NOTES

elephant motif is taken from the unofficial emblem of the Squadron's first operational aircraft, used in 1934. It was the Martinsyde G100, nicknamed 'Elephant'. The elephant also appropriately refers to the squadron's long connection to India.
2. wars/wwtwo/burma_campaign_01.shtml
3. Ted became a very active lifelong member of the South African Air Force Association. One of the Association's objectives was to establish and administer bursaries, grants and loans to provide for the education, maintenance and well-being of the dependants of ex-members of the SAAF.

Epilogue
1. Ted's medals, from left: Distinguished Flying Cross, 1939–45 Star, Africa Star and clasp, Burma Star, General Service Medal, Defence Medal, Africa Service Medal (Ouma's Gong), Malta George Cross Fiftieth Anniversary Medal, and Rhodesian Service Medal.
2. John Gillespie Magee was a pilot in the Royal Canadian Air Force and a member of 412 Fighter Squadron during the Second World War. He was stationed in the United Kingdom. The poem *High Flight* was inspired by a high altitude test flight he made in a Spitfire in September 1941, and which, on landing, he jotted down on the back of a letter written to his parents. Three months later, on 11 December 1941, he was killed in an air crash.

Bibliography

Acknowledgements
Holland, James, *HEROES: The Greatest Generation and the Second World War*, Harper Perennial, London, 2007.
Schucman, Helen and Thetford, William, *A Course in Miracles*, Viking Publication, California, 1985.

Chapter 1: In at the Deep End
AIR 27/1341 No. 217 Squadron, Operations Record Book.
Catton, Gill, 'The Legend of Don Tilley', article in *Wings Over Africa* magazine, South Africa, 1978.
Hay, Ian, *The Unconquered Isle: The Story of Malta*, Hodder & Stoughton, London, 1943.
Lucas, Laddie, *Malta: The Thorn in Rommel's Side – Six Months that Turned the War*, Stanley Paul & Co, London, 1992.
The National Archives, United Kingdom.

Chapter 2: Ignorance is Bliss?
AIR 27/1341 No. 217 Squadron, Operations Record Book.
Hay, Ian, *The Unconquered Isle: The Story of Malta*, Hodder & Stoughton, London, 1943.
Holland, James, *HEROES: The Greatest Generation and the Second World War*, Harper Perennial, London, 2007.
Holland, James, review of Patrick Bishop's *Bomber Boys: Fighting Back 1940-1945*, Harper Perennial, London, 2007.
The National Archives, United Kingdom.
www.secondworldwarforum.com

Chapter 3: From Whence He Came
Lee, Laurie, *I Can't Stay Long*, Penguin Books, Harmondsworth, 1984.

Chapter 5: Like Father, Like Son
Changuion, Louis, *Pietersburg. Die Eerste Eeu*, Pietersburg Town Council, 1986. Changuion, Louis, *Haenertsburg 100, 1887-1987*, Review Printers, Pietersburg, 1987.
Keller, Helen, *The Open Door*, Doubleday, 1957.
Klein, Harry, *Valley of the Mists*, Howard Timmins, Cape Town, 1972.
Levy, Wally, interview with author, 1998.
Wongtschowski, B.E.H., *Between Woodbush and Wolkber: Googoo Thompson's Story*, private publication, printed by Review Printers, Pietersburg, 1987.

Chapter 7: Bucking the System
Heydenrych, Carey, *Heck! What a Life*, Heydenrych Publications (2nd ed), Cape Town, 1996.
Rogers, Lt Gen R.H.D. 'Bobby', letter to author, 1997.

Chapter 8: Taking Flight
Nesbit, Roy C., *Torpedo Airmen*, William Kimber, London, 1983.
Rogers, Lt Gen R.H.D. 'Bobby', letter to author, 1997.

BIBLIOGRAPHY

Chapter 10: Taking the Controls
Bach, Richard, *Stranger to the Ground*, Dell Publishing, New York, 1990.
Rogers, Lt Gen R.H.D. 'Bobby', letter to author, 1997.

Chapter 11: The Big, Wide World
Cavafy, C.P., *Collected Poems*, Princeton University Press, 1992.
Heydenrych, Carey, *Heck! What a Life*, Heydenrych Publications (2nd ed), Cape Town, 1996.

Chapter 12: Puttin' on the Ritz
Heydenrych, Carey, *Heck! What a Life*, Heydenrych Publications (2nd ed), Cape Town, 1996.

Chapter 13: Dive-Bombing and Boyish Pranks
Shakespeare, William, *The Complete Works of William Shakespeare: Macbeth*, Collins, London, 1970.

Chapter 14: Where There's a Will, There's a Way
Gibbs, Patrick, Wing Commander DSO, DFC and Bar, *Torpedo Leader*, Grub Street Publications, London, 1992.
Lucas, Laddie, *Voices in the Air*, Arrow Books, London, 2003.
Murray, William Hutchinson, *The Scottish Himalayan Expedition*, Dent, London, 1951.
Nesbit, Roy C., *Reported Missing: Lost Airmen of the Second World War*, Pen & Sword Aviation, Barnsley, 2009.
Reilley, Herschell, letter to author, 1997.

Chapter 15: Doing the Flying Cha-Cha-Cha
Cavafy, C.P., *Collected Poems*, Princeton University Press, 1992.
Reilley, Herschell, letter to author, 1997.

Chapter 16: Thumb Twiddling
Cavafy, C.P., *Collected Poems*, Princeton University Press, 1992.
Reilley, Herschell, letter to author, 1997.
Whiston, Eddie, letter to author, 1997.

Chapter 17: Bombay and Biding Time
Glennie-Carr, Jinx, letter to author, 1997.
Reilley, Herschell, letter to author, 1997.

Chapter 18: A Flamer – As Fate Would Have It!
Cavafy, C.P., *Collected Poems*, Princeton University Press, 1992.
Whiston, Eddie, letter to author, 1997.

Chapter 19: The Soft Touch of Home
Cavafy, C.P., *Collected Poems*, Princeton University Press, 1992.

Chapter 20: Burma: 'With Speed To the Stars'?
Cavafy, C.P., *Collected Poems*, Princeton University Press, 1992.
www.bbc.co.uk/history/worldwars/wwtwo/burma_campaign_01.shtml

Epilogue
Bach, Richard, *A Gift of Wings*, Pan Books, London, 1976.
Lee, Laurie, *Two Women*, Penguin Books, Harmondsworth, 1984.

Appendix
Miller, Nathan, *Theodore Roosevelt: A Life*, HarperCollins, New York, 1992.

Index

Abbotsinch, 104-109
Addu Atoll, 176, 180
Adkins, Warrant Officer, 85
Aldis signal lamp, 101
Alessio, Riccardo, vii, ix, 206, 207
Alexandria, 7, 149, 216
Aliwal North, 30
All Saints Chapel of St Mary's Cathedral, 180
Allahabad, 183
Anglo-Boer War, 26, 29, 55, 76
Anti-aircraft,
    balloon, 93
    flak, 18
    fire, 9
Anuradhapura, 162
Arakan Province, 167, 183, 185
Avro Anson aircraft, 87-90, 111

Bach, Richard, 82, 200, 221
Baghdad, 140-1
Bambatha Rebellion, 32-3
Bangalore, 152, 161
Baretta pistol, 120, 126
Bay of Bengal, 187, 191
Bay of Biscay, 112
Beaufighters, 21, 157, 165-8, 183, 185, 187-9, 191-3
Bechuanaland, 23
Belgium, 33
Bengal, 183, 187, 191
Benghazi, 114
Bennett, Sergeant, 67-9
Berlin, Irving, 98
Bishop, Patrick, 12, 20, 220
Blackburn Botha, 107
Bloemfontein, 54, 61, 86
Bombay, 152, 159, 160-1
Bomber Command, 12, 91
Bombing raids on Italian fleet, 6
Botha, General Louis, 107
Bournemouth, 96, 97, 99, 103
Boy Scouts, 42-3
Brindisi, 114

Bristol Aircraft Company, 111
Bristol Beaufort
    Mark I, 1, 3, 6, 9, 10, 13, 101, 102, 109, 111, 112, 114, 115, 127, 135, 149, 155, 157, 159, 160, 165, 169, 214, 218
    Mark II, 140, 145, 153, 214, 218
Bristol Blenheim Bombers, 110
British Imperial Airlines, 112
Brixton Cemetery, 38
Brodziac, Gordon, 67, 73, 214
Bronkhorstspruit, 28
Brown, Ray 'Brownie', vii, 21, 113, 117, 121, 136, 139, 140, 149, 151
Buchan, John, 46-7
Bulawayo, xi, 30, 65, 67-8, 70, 72, 74-7, 79, 84-5, 91, 155, 173, 177-8, 195-7, 215, 218
Bulcher, Bernie, 43
Burma, 156, 166-7, 170, 180-3, 185, 187, 190, 191
    The Burma Campaign, 185

Cairo, 131, 134-7, 139, 140, 146
Caliope, 114
Cameron, Squadron Leader, 188-9
Cant Z. 506B, 117-18, 123-4, 127, 131, 202, 206
Cape Bon, 3
Cape Town, xii, 91, 93, 214, 220-1
Catania Air Base Regia Aeronautica (Sicily), 129
Catton, Gill, 1, 61, 67, 87, 100, 213
Cavafy, C.P., 87, 113, 130, 141, 153, 167, 176, 183
Central Air Force Training Depot (CAFTD), 60
    Air Gunner Training Unit, 60
Ceylon, 3, 8, 109, 111-12, 136, 140, 142, 146, 148-9, 152-8, 161, 163, 165-6, 176, 182-3, 188, 193, 195, 213
Changuion, Professor Louis, viii, 44, 46, 217
Cheerio Halt, 47

INDEX

Cheetah jet, 211
Chidell, Pilot Officer Dereck, 70
Chief Magoeba, 46
Chifari, Pilot Lieutenant, 117-18, 120, 206
Chiringa Airbase RAF, 166-7, 183
Chittagong, 183
Chivenor, 100-103, 203-204
Clarke, Nurse Penny, ii, 171, 193
Clarke family,
   Pop, 76-9
   Bertha (Bee), 75, 77-80, 83-4, 91, 155, 165, 170, 173, 177, 179, 181, 196-8, 200
   Ernie, 78-9
   Vi, 177-8
   Granny, 76-7, 80
Coastal Command, 86, 91
Coldbeck, Harry C., 113-14, 129
College of the Little Flower, 44
Collins, Dan, 45
Collum, Edith, 38, 54
Colombo, 155, 161, 163-5, 167, 170, 172, 176, 182, 191, 193. 213
   35th British General Hospital, 170, 175
Connock Motors, 37
Cowderoy, Dudley, viii, 202, 203
Cranborne, 82
Crimean War, 27

Dakoto, 132-3, 170
Dalzell, Group Captain, 83
Darling, Squadron Leader, 107-108
David, Group Captain, 187
De Kok, D.J. 'Old Kokkie', 45
De Marillac, Raymond, 67, 74, 215
Delarey, General Koos de la Rey, 55
Delarey Regiment, C Company, 55, 57
Derstan, Air Commodore, 174, 182
Dunsmore, Bill, 2, 7, 9, 12, 17, 19, 21, 110, 113, 115, 116, 117, 118, 119, 120, 121, 122, 130, 131
Durban, 27, 38, 50, 52-4, 63, 66, 84-5, 195, 214
   SAAF Electrical and Wireless School, 63
Du Toit, Graham Colonel, viii

East Africa, 175-6
East London, 63, 68
Economic depression, 36, 45, 52
Edwards, Corporal, 171, 193
Eerste Fabrieken, 28, 30
Egypt, 7, 21, 33, 111, 131-2, 137, 142, 146, 214, 216
El Alamein, 18
   Memorial, 214
   War Cemetery, 214
Ellis, Dick, 2, 21, 110
England, xii, 3, 23-4, 27, 31, 76, 91, 96, 102, 104, 110-12, 115, 155, 193, 195
Ennis Mine, 47-9
Evans, Cliff, 61, 67, 70, 214

Far East, 3, 5-6, 8, 109, 111, 140, 148, 180, 204
Farrell, George, 25
Filton Aerodrome, 111
First Operation/shipping strike, 12, 22, 112
First World War, vii, 26, 33, 34
Flight Command, 18, 156, 165, 182
Flying boats, 91
Force 136, 188, 189, 191
Foxall, Warrant Officer, 183
France, 33, 96
Frecci, 114

Gain, Squadron Leader Larry, 109
Ganter, Flight Sergeant, 188
'George' automatic pilot, 3
George, South Africa, 86-9
George Peak, 89, 90
Gerlach, Anton, 52
German East Africa, 33
German South-West Africa, 33
Gerny, Boyd, 167-8, 170-1, 174, 193
Gibbs, Squadron Leader Patrick, 13, 115-16, 127
Gibraltar, 3-7, 111-12
Glennie-Carr, Jinx, vii, 164, 174, 221
Gozo, 122
Gray, Bob, 2, 9, 12, 21, 110
Greece, 113, 206
Greek coast, 19, 115, 117, 201, 214
Greenock Harbour, 93-5, 104, 108
Gulf of Manaar, 152, 161

Habbaniya, 140-1, 149
Haenertsburg, ix, 45-6, 200, 202, 208, 211, 220
Haffenden, S.J., vii, 126
Haffrey, Harry, 160
Haggard, Ryder, 42, 46
Hague, Dick, 66
Hallmark, Squadron Leader, 83
Hamilton, Lord Malcolm Douglas, 69
Harrison, Charlie, 148
Harrison, Squadron Leader, 91
Hartley, Captain, 58, 72, 100
Hartley, Squadron Leader, 58, 72, 100
Harvards, 82, 183, 192
Hatherley, 28-30
Hawker, 202, 204
Hay, Ian, 8, 13, 17-18
Hearn-Phillips, Norman, 202
Hewitt, Reverend, 180
Heydenrych, Carey, vii, 61, 73, 87, 89, 91, 93-7, 100, 103, 214-15, 220-1
Higgins,
    Granny, 27-30
    Pop, 27-30, 216
Hijacking, 124
Hitler, 31
HMS *Bermuda*, Fleet Air Arm Station RAF, 164
Holland, James, vii, ix, xi, 12, 20
Hudson bombers, 100, 102, 155, 214
    anti-sub patrols, 91
Hyams, Louis, 45

Induna Aerodrome, 70-2, 81-3, 88, 110, 177, 196
    Indaba Tree, 83
Irrawaddy River, 183, 185, 188
Ismailia, 148-9
    Great Bitter Lake, 148
Italian fleet, 6-8, 13, 22, 112, 216

Japanese, 156, 185, 186, 188
    attack Ceylon, 109
    attack Pearl Harbor, 93, 95
    fighter aircraft, 110
    forces, 155
    occupation, 183
    onslaught, 155
    ships, 110
    surrender, 192
Jiwani, 149, 150-1
Johannesburg, 26-7, 34, 37-8, 40, 43, 53, 54, 57, 66-7, 77, 91, 155, 177, 180-1, 195-6
Jones family, 102, 105
    Joy, 97
    Wendy, 97
    Ouma, 97, 99
    Pharaoh, 97, 102
Jonker Diamond, 29, 216

Karachi, 142, 145-6, 149, 151, 155
Keller, Helen, 40, 220
Khedive, HMS, 193-4
Khumalo Air Training Base (ITW & SFTS) RAF, 65, 67, 70, 73, 82-4, 85-6
Kimberley, 25-6
    siege of, 26
Kisumu, 176
Klein, Harry, 46, 220
Klerksdorp, 54-6
Knight, Derrick, 109
Koggala, 176, 191
Kuhn, Susan, vii

Lamble, Clarrie, 52
Lander, Commanding Officer Johnny, 173
Ledger, John, 73, 87
Lee, Laurie, 23, 200
Leibbrandt, Robey, 31
Leuchars Training Base, 8, 109
Lever, Jack, 67, 87, 195
Levy, Wally, vii, 42
LG224, 131, 133-4, 142, 146, 149
Lockheed Hudson, 91, 100, 102, 155, 214
Loftus, Colonel Dougie, 136
London, 23, 63, 95-8, 128, 157
    Kew Gardens, 23
    Luftwaffe bombing devastation, 96
    New Year, 97
    Old Victoria Theatre, 24
    The Ritz Hotel, 98
    Waterloo Station, 95-6, 128
Long Tom, 26
Lowman, Ollie, 55
Luftwaffe, 18

INDEX

bombers, 96
   raids routine, 18
   medium bombers, 18
   enemy air raids on Malta, 6
   Ju52, 121
   Ju8, 17
   Me109 fighters, 3
   Me110, 17
   Macchi MC202, 117
Luqa Airbase, 1, 2, 5, 6, 10, 13, 14, 113, 129, 130, 131, 132, 133, 201
Lyari River, 146
Lyneham Aerodrome, 110-11

Madras, 187
Mae Wests, 160
Mafeking, 23, 34, 35-6
Magee, John Gillespie, 209, 219
Magoebaskloof, x, 46, 200-201, 212
Malta, 1-8, 10, 13, 17-21, 111-14, 116, 118, 120-33, 141, 148, 152, 153, 155, 196, 200-204, 206, 213-14, 216, 220
   air raids and the alerts, 13
   bofors guns, 12-13
   Malta George Cross Fiftieth Anniversary Medal, 203
   Malta High Commissioner, 203, 206
   Maltese Embassy, 204
Mampas Kloof, 47
Mandalay, 183
March, Doreen, 163, 174
March, Eric, 163, 191
Marham, 202
Marks, Sammy, 28
Mastrodicasa, 120, 206
Matopos Hills, 85, 197
McCouat, Warrant Officer J.B., 167, 170-1, 174
McSharry, Pilot Officer, 3-4, 7, 21
Mediterranean, 5, 7-9, 10, 14, 115-16, 122, 213-14
Middle East, 13, 91, 111, 142, 145-6, 213, 214
Mingladon, 188, 190-3
Minneriya Airbase RAF, 155-7
Minster, Pilot Officer, 3-4, 7, 13
Mombasa, 176, 180

Monviso, 114
Mooketsi, 47
Morgan, Joe, 61
Morgan, Wally, 61,66-7, 70, 84, 87, 100, 104, 106, 215
Morkel, Michael, ix, 201, 208, 211
Mount Lavinia, 165
Mount Pedro, 161
Mountbatten, Lord Louis, 185
Murray, William Hutchinson, 113
Mussolini, 123
Mustafa, Church of, 17
Myburgh, George, 48

Naccari, Mila, ix
Nairobi, 176, 180
Natal, 30, 33, 49, 53
Natalspruit, 26
Nesbit, Roy C., vii, viii, 5-6, 14-15, 114, 125-6, 128, 172
New Milton, 97, 102
New Zealand, xii, 21, 113, 119, 131
No. 13 Navigation and Reconnaissance War Course, 87
Norse, Dudley, 53
North Africa, 19, 63, 137-8
Ntabazinduna Elementary Flying Training School (ETFS) RAF, 70
Nuwara Eliya, 161
Nylstroom, 41

O'Shea, Zadie, 44
Ohio, 131
   entered Grand Harbour, 131
   famous oil tanker, 131
   saved Malta, 131
   'Santa Maria' Convoy, 204
Operation *Harpoon* from Gibraltar, 5
Operation *Vigorous* from Alexandria, 216
Operational record book 217 Squadron RAF, 6, 10, 13
Osler, Bennie, 217, 37
Ossewabrandwag, 31
Outdshoorn, 31
Outeniqua Mountains, 89, 90
Oxfords, 80, 82, 84-5

Paisley, Scotland, 108

Palestine, 33, 140
Pantelleria, 3, 13
Parucatti, Mr, 42
Paul, Group Captain, 170-1
Pearl Harbor, 93, 95
Persian Gulf, 141-2
Phillips, Sir Lionel, 47
Pietersburg Hoërskool, 45
Pietersburg, 39-47, 50, 67, 217, 220
Polonnaruwa, 162-3
Port Elizabeth, 23, 25
Port of Spain, 93-4
Port Shepstone, 50-2
Portreath Aerodrome, 3, 111
Potchefstroom, 177
Potgietersrus, 42
Pretoria, 28, 31, 33-6, 40-1, 61, 63, 65-8, 85, 196, 216
Prevesa, 117-18, 124, 127, 206
Pride, Elizabeth Anne, 25
Primrose, 34, 67

Queenstown, 27, 31

Rangoon, 183, 185, 188-9, 191, 193
Ratmalana, 152, 155, 170, 172, 174, 183
Red Cross, 118, 177-8
Rees, Taffy, 177
Regia Aeronautica, 129, 206
Reilley, Herschell, vii, 7, 122, 130-1, 134-6, 140, 147, 150-1, 156-7, 159, 164, 194, 218
Reilley, LAC, 171
Rhodesian Military Hospital, 177
Roberts Heights (later Voortrekkerhoogtre, now Tswane Heights), 59
   AMA depot SAAF, 59
Robertson, Major Frank, ii, 170, 193, 213
Rogers, Lieutenant General R.H.D. (Bob, 'Bobby'), Chief of the South African Air Force, vii, 60-3, 67, 69, 70, 79, 81, 82, 215
Rogers, William, 180
Rommel, Field Marshall Erwin, 18, 114, 216, 220
Royal Air Force, 64, 83-4, 86-7, 101, 104, 109, 113, 140, 142, 148-9, 164-5, 172, 200, 203, 204, 213-14

No. 11 Squadron, 110
No. 27 Squadron, 183, 202, 203, 204, 205, 218
No. 27 Squadron Insignia, 183
No. 39 Squadron, 13, 21, 214
No. 86 Squadron, 21, 201
No. 209 Squadron, 191
No. 211 Squadron, 166-7, 170, 182-3
No. 217 Squadron, 6, 10, 13, 20, 109, 131, 155, 201, 204, 220
Royal family, 196
   King George VI, 196
   Queen Mary, 198
Royal New Zealand Air Force (RNZAF), 113

Saliermo, Pilot Lieutenant, 117-18
Saline bath burn treatment, 171-3
Salisbury, 82, 198
Saloman, Leo, 45
Scilly Isles, 111
Scotland, 8, 93, 109, 111
Second World War, viii, 31, 57, 219, 220-1
   generation of, vii
   veterans of, xi
Seeber, Monica, ix
Sekwana, Pete, 208
Sentinel L5, 187, 191-2
Seychelles, 176, 180
Shakespeare, William, 100
Shallufa, 139, 147-8
Shandur, 146, 149
Sharjah, 142-5, 149
Sheldon, Hugh, 61, 67, 87, 100, 191, 195
Siam, 185, 188
Sicily, 7, 121-2, 129
Singapore, 193
Sittang River, 185, 188
Skyjack, iii, viii, xi, 129-30, 204, 206
Slim, Field Marshall Sir William, 183
Smuts, General Jan, 31
South African Air Force, ii, viii, 57-9, 63, 118, 170, 175, 196, 211, 215, 217
South African Air Force Association, xi, 60, 211, 219
South African Railways, 30-1, 34, 36, 54, 58

INDEX

South East Asia Command, 172
Southern Rhodesia, 64-5, 67, 87, 178, 196, 200
Spitfire, 18, 113-14, 119, 120, 122-3, 129, 131, 141, 168-9, 191-3, 219
St Paul's Bay, 122-7, 129
    Hospital, 122
    St Paul's shipwreck 60AD, 127, 129
Stamford Hill, 84-5
Stanford, Andrée, 200
Stanford Farm, x, 200-201, 207, 212
Stormjaers (Brown Shirts), 31
Strathnaver, HMS, 91, 92, 94
Strever,
    Albert, 30-1, 54
    Alfred, 23-4, 38
    Andrée, vii, x, xii, 201, 203, 208
    Carolyn Mary, 196-7
    Charlie, 30
    Edward, 30-1
    Gary, ix
    Gertrude, 30
    Gustave, 30
    Michael, ix
    Sheila, 42-3, 46, 48, 50-3, 63, 181
    Shirley, ix
    William Henry 'Pop' (Bill), 26, 30, 33, 36, 39, 40, 42, 45, 58, 138
Strydpoort Mountains, 47
Stutterheim, 27
Suez Canal, 146-148
Swartkop, 85

Ta' Qali, 21, 132, 133
Taranto, 119, 121, 202
Taylor-Smith, Dr, 49, 217
Tempe Camp, 33, 61, 63
    Ground School, 61
Thaba Nyaka, 77
    Serten Mine, 77
Thompson, Googoo, 47, 220
Thompson, Potty, 200
Tiger Moths, 70-1, 81-2
Tilley, Lieutenant Colonel Don, xi, 1, 61, 64, 67, 68, 87, 100, 104, 106-108, 209, 213, 220

Torpedo, viii, 1, 5, 6, 8, 9, 11, 12, 21, 22, 65, 91, 100, 101, 105, 106, 107, 108, 109, 114, 115, 116, 127, 166, 185, 187, 201, 213, 220, 221
Torpedo bombers, 12, 91, 100, 107-108, 201, 213
Toungoo, 191-2
Transvaal, 29, 39, 50, 110, 200
Trichinopoly, 164
Trinidad, 93
Tzaneen, 46

Umzimkulu River, 50

V Force, 187-8, 190-2
Valetta, 15, 18, 204, 214
Van Heerde, Dean, ix
Van Heerde, Lauren, vii
Vavuniya Airbase RAF, 156-9, 161, 163-5, 167, 169, 170, 172-3, 193
Venter, General 'Boetie', 175
Verey Pistol, 10, 117
Villabrod, Father 'Willie', ix

Wall Street Crash, 45
Wanderers Club, 54
Warmbaths, 41, 85
Warren, Major, 188
Wattington, Hugh J., vii
Whiston, Eddie, vii, 157, 172, 174
Wilkinson, John 'Wilkie', 7, 21, 113-14, 117-19, 120, 125, 127, 136, 149
Williams, Mrs J., viii
Wilson, Willy, 82, 84, 87, 214
Witkoppens, 43
Wolf, Major, 47
Woodward family, 25, 27
    Elizabeth 'Lizzie', 23-7, 31, 38
    Esther Pride, 23, 34
    Katie, 23-7, 49, 50, 179, 180
    May, 23, 25, 38, 41, 50

Yelahanka, 152
Ysterberg, 42
Yugoslavia, 118

Zayatkwin, 192-3

Tracing Your Family History?

Read Your Family HISTORY
ESSENTIAL ADVICE FROM THE EXPERTS

FREE COPY!

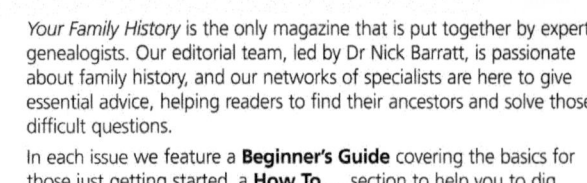

Your Family History is the only magazine that is put together by expert genealogists. Our editorial team, led by Dr Nick Barratt, is passionate about family history, and our networks of specialists are here to give essential advice, helping readers to find their ancestors and solve those difficult questions.

In each issue we feature a **Beginner's Guide** covering the basics for those just getting started, a **How To** ... section to help you to dig deeper into your family tree and the opportunity to **Ask The Experts** about your tricky research problems. We also include a **Spotlight** on a different county each month and a **What's On** guide to the best family history courses and events, plus much more.

Receive a free copy of *Your Family History* magazine and gain essential advice and all the latest news. To request a free copy of a recent back issue, simply e-mail your name and address to marketing@your-familyhistory.com or call 01226 734302*.

Your Family History is in all good newsagents and also available on subscription for six or twelve issues. For more details on how to take out a subscription, call 01778 392013 or visit **www.your-familyhistory.co.uk**.

Alternatively read issue 31 online completely free using this QR code

*Free copy is restricted to one per household and available while stocks last.

www.your-familyhistory.com